DEVELOPMENT PLANNING IN BANGLADESH: A STUDY IN POLITICAL ECONOMY

NURUL ISLAM

Development Planning in Bangladesh

A Study in Political Economy

St. Martin's Press · New York

 For information, write:
St. Martin's Press, Inc., 175 Fifth Avenue, New York, N.Y. 10010
Printed in Great Britain
Library of Congress Catalog Card Number: 77-77354
ISBN: 0-312-19694-6
First published in the United States of America in 1977

TO

CHUNI

Contents

Preface

Bangladesh embarked upon planning for social and economic development immediately after independence within the framework of what the political leadership chose to call socialism and democracy. She had no planning machinery to start with. She not only had a young and inexperienced administration, but also a political leadership which after many years in opposition—first in its struggle for autonomy and then for independence of Bangladesh, including prison cells in Pakistan—came to occupy the seat of the government. The country had gone through the traumatic experience of a war of liberation with consequent dislocation in the economy and in the institutional structure. She had high hopes and great expectations that political independence would bring opportunities to shape the economy of Bangladesh according to her own priorities and within her own perspective, free of the colonial exploitation of the Pakistani regime.

The planning machinery was to be headed by academic economists and manned by professionals drawn from various sectors of the economy but only a few from the traditional civil service and bureaucracy. The academic economists at the helm of the Planning Commission who had little direct experience in government administration, even though one or two had close acquaintance, had confidence in their ability to muster the administrative rules and procedures in time; they saw their assignment as a challenge and an opportunity for committed economists to test their skill in the art of development planning and policy-making.

This book is an attempt to explain how the planning machinery came into action; what problems it faced in terms of application of economic tools of planning and analysis; how it functioned in relation to the rest of the governmental machinery and decision-making process. As the planning process unfolded and economic policy decisions were being taken, the tools of economic analysis and the rationale of policy prescriptions were tested against and confronted with social, political, administrative and institutional realities. We seek to show how, in Bangladesh, planning as a political process evolved and how the Plan—not just a technical-economic but also a socio-political document—took form and functioned.

The book covers mainly the period between the beginning of

1972 when Bangladesh emerged as an independent state and the middle of 1975 by which time the author had already left the government. References beyond this period are few and selective and are used to carry forward the analysis of only certain specific policies of the earlier period. There was a certain uniqueness about the nature and characteristics of this short initial period, as well as about the policy issues which were debated. This was the chief justification for a critical review and analysis of economic planning and policies of this period.

Since Bangladesh was a new country, many issues which were long resolved and for which solutions were already found in an old, established society came up for resolution and decision. Old procedures, traditions and institutions were questioned in a way in which they would not have been, if Bangladesh was not trying to establish them for the first time. Moreover, poverty aggravated by the dislocation of the war of liberation and its aftermath required simultaneous action on all fronts in order to quicken the pace of reconstruction and development within a short period of time. Accordingly, the range of issues and questions which were raised and debated, and on which decisions and actions were required was much wider than under the normal circumstances of a developing country. These three years were very much "packed" with policies and actions and were, therefore, very eventful from the point of view of a policy-maker. Also, the nature of many of the problems and issues which were debated were so fundamental that they were likely to recur in the future with the shifting relative strength of different interest groups as well as in response to changes in the socio-political circumstances. The book does not pretend to cover all the policy issues of this period; it includes only a few selected issues from among many which were matters of concern for the policy-makers. Naturally the selection of issues was somewhat arbitrary; however, they were intended to be those which, in the author's opinion, were likely to illustrate most clearly and vividly the interplay of economic and non-economic factors. They were also thought to be of general interest elsewhere in the developing world.

The author owes a great debt of gratitude to Dr. J. Faaland, Director, Chr. Michelsen Institute, Bergen, Norway, and Professor J. R. Parkinson, Professor of Economics, Nottingham University, England, themselves authors of a perceptive book on Bangladesh as a test case for development. They spent a great deal of time and effort in going through the entire manuscript in various stages of its preparation and made innumerable suggestions for improvement on the framework of analysis, logic of argument and style and language. They were a constant source of both encouragement and criticism throughout the writing of this book.

The book was written when the author was a Research Fellow of the International Development Research Centre, Ottawa, Canada. Without its generous financial assistance, which included not only the author's stipend but also research, secretarial and travel expenses, it would have been impossible to undertake the writing of this book. The author most gratefully acknowledges the patience of the Centre and its sympathetic President, Dr. David Hopper, in generously extending the time period beyond the original schedule for the completion of the book. The Centre, however, is in no way responsible for the contents and presentation of the book. The author alone bears the entire responsibility for the views expressed and facts presented. Last, but not least, the author acknowledges his indebtedness to Queen Elizabeth House, Oxford, which provided the physical facilities for work and research, and to St. Antony's College, Oxford, for allowing the author the facility of a Visiting Fellow at the college during the years 1975 and 1976.

The experience of the political economy of planning and development which is related and analysed in this book was gained during the difficult and turbulent years the author spent as the Deputy Chairman of the Planning Commission, Government of Bangladesh, from the beginning of 1972 to the early months of 1975. It would be only proper to acknowledge with humility what the author learnt, for better or for worse, from his colleagues in the Planning Commission, not only Professors M. Hossain, R. Sobhan and A. Rahman, who joined him as the original members of the Planning Commission, but also the many other colleagues drawn from various disciplines and fields of expertise who worked in the Commission. The contribution to the author's process of learning made by the members of the administration, civil service and the political leadership was, of course, the greatest, because the book is, after all, about them and their relationships with economic policy-making. Differences or similarities in attitudes and methods notwithstanding, the author feels a deep sense of gratitude towards them.

Oxford
October 1976

NURUL ISLAM

Introduction

Development planning and policies encompass not only economic but also socio-political and administrative factors. They interact on and reinforce each other. Economic development and policies produce differential impact on various socio-economic classes or interest groups or on different regions of a country. Again, various interest groups or socio-economic classes participate with varying degrees of influence and power in the political process and hence in the decision-making process for economic issues. Every government has non-economic objectives of various kinds which compete with or facilitate the achievement of economic objectives; they influence economic strategies and policies.

The interplay of economic and non-economic factors shaped the course of events and the evolution of economic policy-making in Bangladesh during the period under review, 1972–75. Both economic and non-economic objectives as well as the need to respond to the competing claims of divergent pressure groups and socio-economic classes guided and influenced policies and events in Bangladesh.

Independence raised high expectations but the prevailing poverty was overwhelming and the scarcity of resources was acute. Disruptions caused by war and its aftermath added to economic distress. Faced with the desperate need to escape from the poverty trap, one could have argued that single-minded pursuit of economic growth at the sacrifice of all other objectives and constraints was expected of Bangladesh. Yet there were political and social considerations at work such as the distribution and exercise of political power, as well as the distribution of resources between sectors of the economy and of costs and benefits of development between various interest groups. To argue that such considerations should have been subordinated to the objective of augmenting resources for development and increasing aggregate output is to claim that poverty makes socio-political considerations or factors less urgent or their interplay with economic factors less important. This was certainly not the case in Bangladesh. The historical circumstances which led to the emergence of an independent state of Bangladesh, the nature and composition of the various pressure or interest groups in the society and the character of the political leadership determined the intensity and pattern of interaction between economic and political considerations in development planning in Bangladesh.

The ruling party and political leadership in Bangladesh had acquired its pre-eminent role primarily as a nationalist party, being in the vanguard in the emergence of Bangali nationalism in the 1950s and 1960s. Its main claim to popularity and the basis of its mass appeal was the promise of emancipation from the economic exploitation and political domination of Pakistan.[1] Consequently the party programmes, pronouncements and election manifestos before independence had given prominence to economic development problems and policies. This continued in the post-liberation period. While recognising the formal commitment to economic development as a high priority objective, two qualifications need to be kept in mind. One was that in the years immediately following independence a great deal of time and attention of the political leadership was absorbed by such issues as the restoration of law and order, the making of a new constitution, holding general elections, and securing diplomatic recognition and establishing working relationships with the various states of the world. These tasks strained the limited abilities of young administration and political leadership. The pursuit of a few of the non-economic objectives had positive implications for economic development. For example, the recovery by the government of arms and weapons, obtained by miscellaneous private persons and groups during the war of liberation and held illegally afterwards, and the establishment of law and order were important for restoring public confidence in the security of life and property. This was obviously a precondition for the revival of economic activity from the dislocation caused by the war. Similarly, the establishment of friendly relations with the rest of the world, especially the developed world, was an essential prerequisite for the expansion of trade and for securing badly needed foreign economic assistance.

The second qualification to the formal commitment of the political leadership to economic development was that the statements of objectives and policies were of a very general nature; even when they referred to specific programmes and projects, they were often not the result of systematic and careful examination by the political leadership. During the 1960s the preoccupation was with policies and strategies which would enlarge the share of former East Pakistan (Bangladesh) in available resources and in the decision-making process in Pakistan. By the time of independence there was no set of widely-shared convictions or a consensus within the party or among the leaders of opinion at large, either about the character and nature of the future economic system of independent Bangladesh or about the strategies and policies of development. Whatever analysis was undertaken on the objectives and policies of development planning in Bangladesh was done in the context of successive Five-Year Plans of Pakistan, Moreover. socio-political circumstances as well as economic policy choices were very different in Bangladesh compared to what they were before independence. As a consequence, the economic principles of the new constitution formulated in 1972 or the economic components of the election manifesto in 1973 were in the nature of broad declarations of intent, which were not formulated as a result of serious and extensive debates and dialogues among the competitive groups and divergent opinions in society; on the contrary they most often marked the beginning of debates, examination, scrutiny and analysis as years went by. Even the formulation of the Five-Year Plan did not become nor was it treated by the political leadership as an occasion for hard policy choices on specific issues; it was treated more as a guideline or a framework for decision-making in subsequent stages, specifically at the time of preparation of Annual Plans.

In evaluating the planning process and performance of Bangladesh it is necessary to examine the nature and composition of the various socio-economic classes or groups which directly or indirectly influenced the economic policies and strategies through their control of or influence on the political process and machinery. The important interest groups were the affluent surplus farmers, trading and marketing intermediaries (mercantile class) engaged in domestic and foreign trade, small-scale entrepreneurs who were aspiring to develop into medium and large-scale industrialists; they played a pre-eminent role in the economic life of the country and indirectly in the political process. In addition, the industrial trade unions were a numerically small but organised pressure group, exercising influence far in excess of their numbers. The students were a separate category; they represented a wide cross-section from among the various interest groups but contained within themselves a wide spectrum of ideological factions. These were the main interest groups from which were also drawn the members of the public service, including the army, the professionals and the intelligentsia. The ruling political leadership, therefore, was basically respresentative of these important interest groups.[2] They expected that political independence would throw open to them greatly expanded economic opportunities in trade, industry and administration, which were so long denied to them by the Pakistanis.

As the nationalist struggle intensified the leading party, i.e. the Awami League, had to broaden its support amongst the masses. To obtain a universal mass following, it was not enough nor even crucial to hold out hopes of restoration of democracy, majority rule or for that matter, autonomy for East Pakistan; the party programme had to hold out the promise of economic uplifting of the masses. It promised improvement for the lot of the rice farmers, jute growers and sharecroppers; control of floods which brought misery to millions and washed away the results of these efforts and investments. and reduction of tax burdens on the small and medium farmers. Public control or ownership of large-scale industrial and financial fortunes, which were owned by Pakistanis who were seen to be in support of Pakistan's political–military domination of Bangladesh, was soon adopted as components of the party programme. To a considerable extent the struggle against vested interests and the pursuit of egalitarian policies coincided with the struggle against Pakistani domination.

In an attempt to broaden its support and increase its following among the masses, the leading party had not only to adopt a mass-oriented programme but also to draw increasingly on the support of students and workers who were valuable allies in mobilising mass support and organising election campaigns and demonstrations. By the 1960s the radical parties had made important inroads and gained support among the students and workers. The wider the net was cast among the students and workers, the greater was the need for introducing radical programmes in the party platform. They constituted the most powerful source of radicalisation within the party. Moreover, the leading party, in order to deflect popular support from the radical opposition parties found it necessary to introduce socialist programmes within its platform. The liberation war, in which the students, young peasants and workers had played a leading role, further raised the radical aspirations amongst the population in general and these groups in particular.

As the struggle for autonomy and independence gained momentum, all classes of society and advocates of political ideologies of the left as well as of the right either joined the party or supported it in its struggle for the realisation of the nationalist aspirations of Bangladesh. By the time independence was achieved, the ruling party had a universal following among all elements in society. The conservative and the radical forces were contending for supremacy within the party. The leadership of the party at the apex stood in the middle and appeared not to have firm convictions of its own in either direction. It was willing to make advances towards socialistic and egalitarian policies provided there were not serious challenges by the conservative groups or vested interests. The leadership of the party opted for a continuous compromise, taking one step in one direction offset by another in the opposite direction.[3] The leadership was in favour of keeping its options open. It could not go against the socio-economic groups which were dominant in the economy and in the organisational structure of the party. It feared that any curtailment in the power and privileges of these groups which they could not be persuaded to accept through the force of enlightened self-interest would alienate them.

It was in this socio-political context that the economic policies and programmes in the years from 1972 to 1975 needed to be evaluated and judged. This was the setting for the establishment of the Planning Commission in early 1972 and the Five-Year Plan which was launched in 1973. The formulation of the Five-Year Plan was one of the few areas of economic policy-making to which the government publicly committed itself in its election manifesto in 1970 as well as in subsequent public pronouncements. The introduction of a full-scale planning process involved a number of tasks: a planning machinery manned by highly qualified persons had to be organised and its role defined; its relationship to the bureaucracy and the permanent civil service, and to the political leadership had to be delineated. This two-fold relationship was crucial in the determination of the effectiveness of the Planning Commission.

The Planning Commission was also formally vested with the responsibility of assisting the preparation of development projects by the other ministries. It was responsible for technical and economic appraisal of such projects before their inclusion in the Plan as well as for *ex post* evaluation of the performance of the economy and of the progress of project implementation. Moreover, it had the task of negotiation and allocation of foreign economic assistance between sectors and projects. It was to be involved in deliberations on all aspects of economic policy-making, programmes and projects as well as their co-ordination, both short and long run. Thus the Planning Commission was entrusted with wide-ranging functions. Yet the full implications of the planning process were inadequately understood or accepted by the political leadership and by the bureaucracy. At the same time the traditional functions and responsibilities were redistributed between the ministries and agencies, between the ministries and the Planning Commission and between the civil service and the professional cadres.

The fact that the planning process imposed constraints on the freedom of choice of individual ministries was not readily appreciated. Individual ministries made pledges for particular projects and programmes to the constituencies they represented and the party made pledges for specific policies; but in the course of subsequent examination the Plan did not always come out in their favour. Not infrequently, fulfilment of such pledges became a test or a symbol of the credibility or success of politicians; therefore, the constraints imposed by the Plan were found to be inconsistent with their freedom of action as peoples' representatives, even when the planning decisions were collectively taken by the Cabinet and should, therefore, have been accepted by individual ministries once they were taken. However, increasingly a trend developed towards the centralisation of decision-making in the hands of the Prime Minister, reducing their sense of participation.

Although the Plan was a socio-political document, there was no public debate or discussion about its objectives, priorities and strategies. It was never presented to the Parliament for debate or discussion. Nor was it exhaustively debated by the Cabinet. The leadership of the Planning Commission was in the hands of technocrats who saw their role primarily as economic advisers. There was no effective and continuous dialogue between them and the political leadership; the political leaders were not closely associated with the deliberations of the planning machinery itself and therefore, were not fully aware of the constraints or the possibilities with which the planners were faced. On the other hand, the planners were not always fully aware of, and, even when they were aware did not fully share, the hopes and aspirations of the political leadership; nor did they share or fully appreciate the politicians' criteria in judging the feasibility of strategies and policies.

There was a lack of effective co-operation with the members of the civil service and the bureaucracy; the bureaucracy and administrative ministries were not eager to accept the procedures of planning or to grant the Planning Commission its leading role in the process of economic policy-making. A Planning Commission dominated by professional experts and headed by academic economists and not by civil servants was already a break with past tradition. Its advocacy that the government should abandon the concept of an "elite" civil service as the centrepiece of the governmental machinery and that it should draw upon a wider spectrum of expertise and experience available in the country made it an alien and unfriendly organisation in the eyes of the established civil service. The political authorities were indecisive about the desirability or otherwise of an elite civil service; they were willing neither to confirm its former pre-eminence and authority nor to supplant it by a suitable alternative system. Rather they tinkered with the existing administrative structure, procedures and rules which resulted in a loss of motivation and confidence without compensating benefits. The reluctance of the Planning Commission to accept uncritically the established rules and procedures, its scant regard for the status quo and precedents and its eagerness to force the pace of change, while a new set of rules of business and procedures were not rapidly formulated, generated tensions. No less important were the differences of opinion, sometimes strong, between the two groups on policies and institutional changes. The civil service was not ideologically or politically neutral. The socio-political changes in Bangladesh had eroded the concept of neutrality.

The Five-Year Plan was formulated on the basic assumption of a mixed economy in a state of transition to a fully socialist system. It overlooked the fact that the political leadership had no firm convictions in this respect; it was not unanimous as to whether a mixed economy would be a transitional stage or whether it was going to be the permanent long-term pattern of the economic system. Moreover, the choices on the size and priorities of the Plan were made by the Planning Commission itself on the basis of its own interpretations of the preferences of the political authorities, as revealed in public pronouncements and in discussions of a general nature with the highest political leadership. The macro-economic implications of the policy recommendations of the Plan were not subject to serious scrutiny in any political forum. The intersectoral allocations of the Plan were, however, matters of intensive and exhaustive discussions with functional and administrative ministries; they differed widely with respect to the preparation of projects and programmes, analysis of sectoral development programmes and priorities, and ability to present and argue their case for allocation of resources.

A simple consistency model, i.e. input-output model, was used by the Planning Commission as an aid to its planning exercises. It helped to work out the broad magnitudes of savings—investment gaps as well as the balance-of-payments gaps resulting from alternative projections abut the increase in per capita income, consumption and final demand. The quantitative application of the model suffered from a serious lack of adequate data, uncertainty about future changes in technical input–output coefficients and changes in relative prices. The knowledge about investment alternatives was quite inadequate in a few sectors due to the lack partly of adequate information and partly of skilled management for undertaking necessary analysis. Provisions for flexibility and future changes in sectoral allocations were made especially in those sectors like physical infrastructure, health, education and family planning, etc,. where results of studies already underway about investment alternatives, including appropriate technology and institutional framework, were expected during the Plan period.

The model, however, revealed the need for and provided a stimulus to the collection of data and it helped establish the identification and preparation of projects; it helped establish criteria and procedures for the economic appraisal of projects. The process of formulation of the model and its application to the preparation of the Plan was an educational one; it forced or stimulated the ministries to think of inter-sectoral linkages, to examine the relationship of their programmes with those in other sectors and to scrutinise policy questions in their sectors in the broader context of the economy as a whole. The Plan required, though not very successfully, individual ministries or sectors to evaluate the alternative ways or techniques of achieving their plans and programmes and to stablish critieria or a rationale of choice between alternatives.

There were administrative and political constraints on the determination of inter-sectoral allocations strictly on the basis of economic criteria alone. For example, a significant constraint on the freedom of choice was the backlog of ongoing projects. In the light of new costs and demand conditions there was a need for re-examination of many ongoing projects. There was the possibility of scrapping, modifying, postponing or mothballing a few of them. However, freedom of choice in this respect was limited by two factors. One was that many such projects were financed by foreign aid and any postponement and modification of such projects would require agreement of the donors. Secondly, ongoing projects created vested interests of their own. Those who were employed on such projects, i.e. managers, supervisors and the labour force, however small in size, constituted pressure groups in favour of their continuance; otherwise they faced either unemployment or relocation to other projects which involved costs of adjustment—a politically unpopular act. Another constraint on the choice of projects consisted of a few large-scale projects, many of which became politically important in the pre-independence days as symbols of the aspirations of Bangladesh (former East Pakistan). Even though they required re-examination in the changed circumstances of market, costs and sources of supply and demand, their postponement or abandonment required politically difficult decisions, implying going back in some cases on electoral pledges carried over from the pre-independence days.

The most politically delicate choice was the relative dependence of the Plan on foreign aid; with the oft-repeated emphasis on self-reliance and socialism, heavy dependence on foreign aid, particularly that from the non-socialist countries, especially the big powers, created a political dilemma for Bangladesh. There was the choice, on the one hand, of severe austerity; the brunt of such austerity would have fallen on the already impoverished masses and this could have been made partially politically palatable, by a universally accepted regime of egalitarian consumption standards; this in turn would have required a degree of ideological motivation, not discernible in post-independence Bangladesh. On the other hand, there was the option of seeking a large inflow of foreign aid from the rich, powerful nations, which brought with it certain restrictions on the freedom of action in political and economic decision-making. Bangladesh opted for larger foreign aid.

Implementation of the Plan brought in its wake other complications, uncertainties and contradictions. There were severe problems of coordination between numerous ministries and executive agencies engaged in policy-making as well as in implementation. There were serious obstacles in the way of downward delegation of authority and responsibility and in the way of decentralisation to the local government institutions. The co-ordination of the activities of local officials of the national government working on local development programmes was far from satisfactory. There was great delay rebuilding the local government institutions disrupted during the war of liberation and establishing new ones. There was ambivalence in the attitudes of both politicians and administrators in the national government towards decentralisation, in spite of the strong rationale for a policy of devolution. The bureaucrats of the national government were averse to a loss of power and authority to the local officers; the politicians operating at the national level were worried that the politicians in local governments would compete with them and secure access to resources for the distribution of patronage; the party in power in the national government was worried about the opposition party or parties seizing control of local governments and using the authority and resources for increasing their popular support.

Two most important obstacles in the way of successful implementation of the Plan deserve special mention. One was the impact of unforeseen factors on the projections of the Plan and on its implementation; the other was the deficiencies in management and organisation. Admittedly, one of the principal characteristics of underdevelopment was the deficiency in organisational and managerial ability. It was not merely the need to establish appropriate institutions; it was also a question of attitudes, systems of rewards and incentives, habits of work, skills and technical knowledge and procedures and systems of administration. The impact of organisational and managerial deficiencies was pervasive, ranging from under-utilisation of capacity to lack of repair and maintenance. There was a vicious circle; the greater the scarcity of capital resources, the greater the need to gain from more efficient and economical use of scarce capital. But this did not happen in Bangladesh where scarcity of resources was compounded by organisational and managerial deficiencies. Worse still, efforts to offset low yields from existing capital equipment caused by organisational and institutional weakness led to demands for further additions to the capital stock. Faced with the need for quick increases in output, the ministries often found it more convenient to generate pressures for larger capital expenditue rather than to carry out the more painstaking and time-consuming task of improvement in efficiency.

Annual Plans were the instrument for the implementation of the Five-Year Plan. The broad outlines of strategies and policies enunciated in the Five-Year Plan became matters of hard choice and policy decisions in the course of the Annual Plans. They specified, appraised and evaluated projects and programmes for inclusion in the annual development programme; they also evaluated the progress and performance of the economy and made concrete recommendations for improvements or changes in policies and institutions. The preparation of the annual budget in which both the Planning Commission and the finance ministry worked closely together involved projecting development and non-development resources and expenditures, and making specific policy recommendations on fiscal and credit policies. The two most important instruments or aspects of this were foreign exchange budgeting and food budgeting. Both foreign exchange and food were very scarce and had to be used efficiently and distributed equitably among competing users. Both were undervalued in relation to what might have been an appropriate price to bring demand down to the level of very limited supply. The public system of distributing and rationing foodgrains covered all classes of population in the four largest urban areas, all government employees throughout the country and only residually the low-income groups in the rural areas. The rest of the population purchased in the free market and bore the brunt of fluctuations in the supply of foodgrains from domestic production and imports. Food planning was mainly, if not entirely, restricted to negotiations for food aid and to purchase of imports to meet the estimated gap between consumption requirements. All imports were channelled through the rationing system. Planning purchases in a highly volatile and uncertain world market for foodgrains included arrangements for shipping, since Bangladesh had no domestic fleet of her own and had to depend upon the uncertain availability of foreign ships and to pay high and fluctuating freight charges. After domestic procurement was introduced from 1973/74, food budgeting also included planning and implementation of domestic procurement of foodgrains from the marketable surplus produced at home. Fixing prices for domestic procurement and for sales under the rationing system posed difficult policy choices; the concern for equity by keeping the price of a basic item low had to be matched against the need to ensure a remunerative price to stimulate increased production.

Foreign exchange budgeting involved making an annual projection of foreign exchange resources from all sources and allocating what was available between the public and the private sectors and between competing users in each sector. The administrative mechanism of exchange budgeting and controls was made more complex by the multiplicity of sources of foreign exchange, often restricted in use, like earnings from barter trade agreements and foreign aid tied to projects and commodities by sources of purchase. Between 1972/73 and 1975/76 the government raised the prices of both food and, by devaluation, of foreign exchange distributed under the system of administrative controls. The increase in the price of rationed foodgrains was in response to a rise in import prices and the need to avoid a corresponding increase in food subsidies. In the case of foreign exchange, devaluation was intended to restore the competitiveness of exports, eroded by continuing inflation at home, and to prevent heavy losses in the jute manufacturing industry, the principal export industry. In both instances it was hoped that the smaller the divergence between free market and controlled prices, the lower would be the incentive to carry on and gain from the illegal transactions.

The Bangladesh economy was heavily dependent both on agriculture and foreign trade. Fluctuations and uncertainties arising from variations in weather conditions and world trade had a serious impact on the planning process and the implementation of Annual Plans. The inadequacy of data, which weighed heavily on the preparation of the Five-Year Plan, was of greater consequence in annual planning because more detailed and up-to-date statistics were needed for short-term planning and forecasting, The methods of forecasting balance of payments and food production left much to be desired from the point of view of technical adequacy; there was a compulsion to err on the side of optimistic forecasts, partly in view of the political need to demonstrate that better results were expected in the future, and partly because offers of foreign assistance were thought to be dependent upon an improvement in economic performance from year to year. From the point of view of the policy-makers in Bangladesh, bad economic performance or the frustration of plan targets due to an adverse turn in external economic circumstances was a double-edged sword. While it could possibly be used partly to cover up deficiences in internal management and organisation, it was not easy to explain to the uninformed public that shortages and high prices were due to the adverse impact of world events. It was politically easier to explain the adverse effects of bad weather and crop failures, But even in this case. it was conceived to be the elementary responsibility of any government to be able to provide food at times of shortage and distress.

Two most important economic developments were very destabilising. One was the succession of bad harvests, two droughts in 1972/73 and floods in 1974, all of which were superimposed on the dislocation of the institutional framework of agricultural production in the aftermath of war. The other was the effects of the world-wide inflation and an adverse movement in the terms of trade. Either was sufficient to create acute economic crisis; the two together exceeded the ability of the economy to withstand adverse fortune and of the administrators to handle it. Under the best of circumstances Bangladesh was a difficult economy to manage and develop; under these circumstances, it created great despair and frustration.

Bangladesh had reserves neither of food nor of foreign exchange resources to cushion the effects of violent changes in prices and availability. Annual plans did in fact become instruments for the design of policies and programmes to reduce the adverse affects of violent swings in weather conditions and world trade. The deterioration in terms of trade widened the gap in the domesic budget as well as that between international receipts and payments. It distorted the relative profitability of growing competing crops such as rice and jute and changed the cost–benefit ratios of projects appraised before the price changes took place. Fluctuations in export earnings were aggravated by fluctuations in the price and availability of imports. Experience in Bangladesh to date indicates that the greater the shortage and the uncertainty of supplies. the greater was the trend towards controls in the belief that the market mechanism would not only distribute income inequitably but would also fail to allocate resources to the economic sectors or groups deserving high social priority. Moreover, allocation of resources in response to widely fluctuating or erratic price signals was likely to be inefficient and result in unnecessary dislocation.

In the years between early 1972 and early 1975, Bangladesh had in fact to plan the availability and use of food and foreign exchange not only from year to year but from quarter to quarter and during the worst crisis of late 1974, caused by a combination of floods and food shortages, from month to month. Even in normal times there were seasonal fluctuations in the availability of foodgrains from domestic production, in view of three rice crops of different sizes being harvested in three different seasons; imports were used to meet the shortages between one crop and another as well as to meet the overall annual shortage. It was, therefore, important that food imports were available in the months of greatest shortage, particularly so in a year of bad harvests as in 1974 when the late arrival of food imports, with no domestic stocks. caused famine and death from starvation. Fluctuations in the volume of imports caused by fluctuations in price and availability of imports caused widespread under-utilisation of capacity and interruption in domestic production; it stopped construction activity, a vital component of investment highly dependent on imported materials, and limited to the winter season.

The attempt to mobilise domestic resources for financing development brought into the open the interplay of economics and politics, and of divergent interest groups. Three sectors of the economy in Bangladesh generated surpluses which potentially could provide sources of capital accumulation. There was the agricultural surplus earned by affluent farmers which was augmented by the application of modern technology; there were the high profits earned in the distributive trade, in the marketing of scarce imports as well as domestic output; and there was the potential surplus of the nationalised sector, mainly the public sector industrial enterprises.

After independence land revenue on holdings below 8 acres in size was abolished; this exempted more than 90% of landholdings from the purview of direct taxation; the collection of the agricultural income tax, which in any case had very limited coverage and a very low rate was postponed for a period of two years after 1973. In spite of increases in the price of agricultural inputs in response to increases in their procurement cost, they were highly subsidised. Prices of agricultural output remained high owing to continued excess of demand over supply. Consequently high profits were earned by surplus farmers, which were either conspicuously consumed or spent on real estate or invested in speculative and quick-yielding trading operations. Even though the benefits of high prices and subsidies accrued predominantly to the rich farmers because of their greater access to and command over resources and inputs, the proposals for increased taxation of agricultural land or income or progressive and rapid elimination of subsidies were resisted. It was argued that high agricultural incomes were not high enough or were not enjoyed for a long enough period or were not widespread among all the surplus farmers. Furthermore, it was felt that the withdrawal of subsidies would slow down the adoption of modern technology and would frustrate the overriding objective of increased food production. However, substantial increases in output were unlikely unless modern technology was adopted very extensively by the smaller farmers, who required more than anything else substantial increases in the supply of credit and an elastic supply of modern inputs through an improved, decentralised system of rural institutions.

Moreover, subsidies on foodgrains distributed under the rationing system compounded the difficulty of mobilising resources for investment. It was not an equitable system since the urban population. including all categories of public sector employees and the industrial workers, irrespective of their incomes, received priority in the distribution of subsidised food whereas the poor in the rural and semi-urban areas were the residual recipients. However, it was politically expedient to give preference to the volatile and organised pressure groups in the urban centres.

Again, high profits in trading activities were not adequately taxed either through high tariffs or exchange rate adjustment or auctioning of licences because it was contended that higher taxes would push prices further up. Affluent farmers or traders constituted powerful vested interests and were not as yet ready to accept an increase in the burden of taxation or surrender the benefits of subsidy.

The utilisation of the under-employed rural labour force provided a potential source of capital accumulation, The absence of political and social organisations at the grassroots level, including the lack of motivation and incentives necessary to ensure an equitable distribution of a highly inelastic supply of wage goods, principally food, among both the old and newly employed labour force and the ownership pattern of the land and the uneven impact of rural projects, potentially eligible for employment of surplus labour rendered this policy alternative an unrealistic one.[4] Faced with scarcity of public sector resources, inflation was often an easy way out as a method of indirect taxation. The traders, speculators and surplus farmers, etc., did not fare badly; fixed income earners and unorganised labour in the rural and urban areas suffered a decline in real income.

The organised industrial workers did obtain wage increases partly on the initiative of the government and partly through collective bargaining, including the threat of violence, until later in the period when strikes were made illegal and repressive measures were taken by the government. However, increases in industrial wages reduced profits or more frequently, increased losses of public sector enterprises. These losses were met by the creation of bank credit and had adverse inflationary consequences. On the other hand, an increase in the incomes of organised industrial workers worsened the inequality between them and the unorganised urban workers in the small-scale sector and the landless labourers. The nationalised sector suffered losses or earned very low profits; it was caught between the stultifying effects of bureaucratic control coupled with managerial and organisational defiencies and a pricing policy which eschewed the exploitation of monopoly profits or scarcity rents. Scarcity premia provided by high market prices of domestically-produced manufactured goods were earned by the trading intermediaries, who escaped any substantial taxation.

The industrial investment policy in Bangladesh became a test case for the socialistic policies and pronouncements of the government. Bangladesh proclaimed itself a socialist state. Partly due to historical circumstances created by industries left behind by

Pakistanis and partly in response to the pre-independence election pledges, about 85% of total industrial assets were placed under public ownership. A ceiling on private investment in any industrial enterprise was fixed at Tk.2.5 million; private foreign investment was allowed only in participation with public enterprise. The debate around the role of the public sector led to a debate about whether Bangladesh was a mixed economy in transition to a fully socialist state or was going to remain permanently as a mixed economy.

There was the question of an appropriate incomes policy for Bangladesh. The principal concern with respect to incomes policy related to the permissible or desirable range of income differentials, mainly if not entirely in the public sector and among the salary and wage earners, including the principles and methods of determination of wages in the public sector enterprises. The political leadership was concerned about this because of the high expectations of social justice and egalitarianism which were raised before and during the liberation war and because of an increasing militancy amongst the industrial workers, remembering promises by the political parties about workers' participation in the profits and the management of industries. The preoccupation with income differentials in the public sector as the cornerstone of incomes policy derived its rationale only in a context in which the public sector was to be progressively expanded so that the predominant part of national income would be generated as wages and salaries with property incomes and profits playing a limited role as a source of income. This concern, if not entirely out of place, certainly did not deserve high priority in the prevailing circumstances of Bangladesh in which outside of large-scale industry, financial institutions and a segment of foreign trade, the private sector generated the bulk of personal income, particularly in agriculture, trade, construction and services, self-employed or otherwise. With a view to reducing income differentials, especially the differences between the highest and lowest incomes, a ceiling was put on salary incomes in the public sector. Very soon, with continuing inflation and a fall in the real income of the highest grades of salary earners, the effect of the ceiling on the motivation, efficiency and propensity towards corruption on the part of the public sector employees was debated. The inequities and inconsistencies became evident when high profits or incomes in the private sector, accompanied by conspicuous consumption, contrasted with the ceiling on the public sector salary incomes, even though the latter were made tax-free.

It was argued that trade unions should abandon their conventional role in public sector enterprises, especially as they were to participate in the management. The structure and level of wages and other benefits were to be decided by negotiations at the national level between the workers' representatives, state enterprises and the government, as part of an overall incomes policy in the public sector, and not by means of collective bargaining either in the enterprises or in the industry. The workers were not enthusiastic about participation in management if it was to be accompanied by a restriction on the scope of collective bargaining. The suggestion that the workers' participation in management, even though in a minority role, should be based on the principle that one enterprise should have only one union, thus eliminating the multiplicity of unions, and that "outside" labour leaders who were not employees of the enterprise could not represent workers in management, was not acceptable to them. They preferred the weapon of collective bargaining and freedom to form a multiplicity of trade unions, competing for the advancement of their interests, in opposition to management. The trade unions straddled both public and private sector enterprises; they could not introduce a dichotomy in their functions, responsibilities and organisational structures for the sake of participation in the management of public sector enterprises.

The debate surrounding the future of private industrial investment turned basically on the question of whether it would be allowed to grow and expand out of the confines of the small-scale industrial sector to the large-scale sector and whether it should be encouraged and assisted by the government to undertake a sizeable share of the total industrial investment in the country. There was a thriving private sector in trade, construction, agriculture and service activities which in a sense constituted the pressure group to enable the aspiring industrialists to achieve a larger role in the industrial sector. They feared that restrictions on private industrial investment would act as a harbinger of further restrictions on private investment and would encourage it in the other sectors and eventually on private ownership of assets or property in general. In addition to the upward revision of the ceiling on investment, they pressed for state patronage, following the example of pre-independence days, in the form of concessional fiscal, financial and foreign exchange facilities and incentives to increase the profitability of private enterprise, so that the latter could expand, leaving the public sector to those activities in which private enterprise was either unwilling or unable to invest.

The controversy regarding the ceiling on the size of investment in industrial units under private ownership became a symbol of struggle between the forces contending for and against the extension of public ownership in the country. On the one hand, there was the question of imbalance between public sector savings and investment as well as between private savings and investment. Could adequate financial institutions and instruments be devised to channel private savings to public investment without disincentive to private savings? It was argued that the ceiling was frustrating private savings and scarce entrepreneurial talents. Private investment opportunities were limited because an adequate number of viable units was not available within this limit, which in any case had been more restricted by inflation since 1972. There was, on the other hand, the counter argument that it was the foreign exchange constraint rather than the ceiling which was the limiting factor on private investment, together with the law and order problem, the militancy of industrial workers and general political uncertainty. Whether investment opportunities were adequate within the limit of the ceiling was an empirical question, which could be verified only if the policy was implemented for a certain period of time; it was too early to resolve the issue in 1974 when the pressure for raising the ceiling gained momentum.

The political leadership, without firm convictions in either direction, was caught in a dilemma and was vacillating. The unpleasant memories of private capitalism in Pakistan with its concentration of economic power, its inequities and its control of the political and economic decision-making process, discouraged the highest leadership of the party from opting unequivocally for private enterprise or for introducing incentives and concessions to enable it to expand and gain in strength. On the other hand, the political leadership was apprehensive that the policy of limiting private investment might limit the growth prospects of the economy. They did attempt to mobilise private savings, however, through raising interest rates on financial assets in 1974 but not adequately in view of rising prices; they did plan for infrastructural facilities for the assistance of entrepreneurs in small-scale industries but did not implement them vigorously.

While controversy about the relative roles of the public versus the private sectors continued, no determined effort was made to improve the working and efficiency of the public sector enterprises. An appropriate institutional framework for the management of public sector enterprises delineating their relationship with the government and ministries was not evolved. The bureaucracy in the government pressed for central direction and control of public enterprises with respect to output, employment, personnel, pricing and purchase and marketing policies. The public sector enterprises struggled for greater autonomy and higher status and authority *vis-à-vis* the bureaucracy. The political leadership was uncertain as to the appropriate degree of autonomy and vacillated. It did not look with favour on any substantial surrender of responsibility to the public corporations; it did not appreciate that day-to-day control over their operations diffused responsibility and divided accountability without guaranteeing efficiency or profitability; it was not confident that *ex post* evaluation of performance could provide a safeguard, if any, against the inefficiency of public enterprises.

By 1974 there were some who earned high profits in trade, speculation and construction and other service activities; they were joined by foreign aid agencies as well as foreign private corporations in advocating the upward revision or elimination of the ceiling on private investment as well as government participation in investment between foreign and domestic private enterprise. At the same time, it was suggested that a declaration of a moratorium on "black money" earned in illegal transactions in violations of trade, exchange and tax regulaions, would discourage export of capital as well as draw funds away from hoarding, towards investment. In July 1974 the ceiling was revised upwards from Tk.2.5 million to Tk.30 million and partnership between foreign and domestic private enterprise was allowed.

Following the military coup and change of government in August 1975, and the successive coups, further changes were made. The government which came into power in November 1975 revised the ceiling to Tk.100 million and moved one step further by allowing joint partnership between domestic private and public sectors. It also announced a moratorium on "black money" provided it was invested in the priority industries. Moreover, liberal fiscal and financial concessions were offered to private investment.

The original investment policy was allowed a short lease of life and that also in the context of highly unstable and difficult economic and political circumstances. It is too early as yet to pronounce judgement on the new policies. It is difficult to predict the unfolding of the interplay of economic and political forces in the future and hence the future pattern of the economic system, including the relative role of public ownership and enterprise.

In Chapter 1 we describe the composition and ideology of the dominant political leadership; the evolution of its thinking on economic system and policy; its political commitment to economic development and planning, and the socio-political perspectives of the first Five-Year Plan. There follows, in Chapters 2 and 3, a discussion of the genesis and structure of the planning machinery and its relationship with the political leadership and with the bureaucracy. We discuss problems of co-ordination of the process of Plan formulation and of policy and programme implementation; we analyse the role of the planning machinery in the task of economic coordination, and with respect to the procedures and criteria for appraisal and evaluation of projects, review of the general performance of the economy and implementation of development projects. A theme throughout these first chapters is the ever present need for compromises in order to seek a second-best solution—a solution that could work and was acceptable to all parties concerned—rather than an ideal theoretical solution.

Chapters 4 and 5 concentrate on the application of tools of economic analysis in economic planning, including the use of models for medium-term planning; they discuss problems the planners faced in terms of inadequacy of data, lack of information about alternative investment opportunities and their costs and benefits, and the need for flexibility and adjustment as new information accumulated and old assumptions changed. They highlight the techniques used and the problems encountered in annual resource budgeting, including foreign exchange budgeting and food budgeting, exercises in planning which were very crucial in a country which was in the grip of severe food and foreign exchange shortages.

Chapters 6 and 7 deal with two selected areas of policy-making which generated a great deal of concern and controversy in policy-making circles as well as amongst the various interest groups dominant in the Bangladesh economy and in its political process. We examine first the problems of domestic resource mobilisation, including taxation, subsidies and mobilisation of surplus in the rural economy, and then the issue of the public versus the private sector in the context of industrial investment policy which became the symbol and centre of debate about the future economic system of Bangladesh. Nowhere else was the intermingling of political, ideological and economic factors more evident than in these areas.

The concluding Chapter 8 ventures to draw a few lessons from the planning experience of Bangladesh which might be relevant in Bangladesh in the future and also might be of some interest to the other developing countries similarly placed. We seek to draw lessons in a few of the areas or issues covered in Chapters 2 to 7. In addition, the concluding chapter speculates on a few aspects of the future pattern of the economic system and policies in Bangladesh.

Notes

1. This was closely related to its espousal of democracy and majority rule in Pakistan as well as its demand for autonomy for East Pakistan (Bangladesh) in economic decision-making and resource allocation.

2. It did not represent the big industrialists and the big landlords, primarily because they were so few of them in Bangladesh. For historical reasons most of the big landlords in Bangladesh were Hindus, who migrated to India in 1947; partly for historical reasons and partly because of policies pursued by the Pakistan government, most of the large industralists were Pakistanis who left at the time of independence of Bangladesh.

3. This was true not only of economic policies but also in terms of

keeping the balance between competing elements in the administrative structure and the party hierarchy, contending for power and authority.

4. Employment of surplus labour at going market wage rates, apart from the organisational problem in the rural areas, would have required an expanded domestic supply of wage goods or a large inflow of food aid, both of which were not likely.

1. Political Factors in Development Planning

(a) The Dominant Political Leadership: its Composition and Ideology

The dominant political force in Bangladesh was represented by the political party known as the Awami League. It was established in the 1950s by the dissident members of the former Muslim League which fought for the creation of Pakistan during the period of British rule. It was a nationalist party and it believed in the democratic form of government. As a reaction against the non-secular religious bias of the Muslim League, it adopted secularism as one of the planks of the party's platform. In terms of its economic philosophy, it was a liberal party with socio-democratic leanings. It assigned an important role to state intervention for curbing the excesses of the private capitalist system and for securing the benefits of a welfare state. Primarily a nationalist party, it represented the revolt of the middle classes of Bangladesh against the economic and political domination of Pakistan. During the Pakistani period, it fought for a larger share in the benefits of economic progress and a larger share of resources for the development of East Pakistan. It wanted greater participation by East Pakistan in the process of political and economic decision-making. By the late 1960s its main platform was the demand for autonomy in the management of the financial and foreign exchange resources of East Pakistan on the basis of what came to be popularly known as the "Six Points Programme of Autonomy".

The Awami League was basically a middle-class party in terms of the composition of the economic interest groups, which provided its main support in the country. They were the surplus farmers, the traders and small industrialists and the lower echelon of the civil service. The number of traders and industrialists was small in former East Pakistan; they suffered from comparative disadvantage in competition with their more established counterparts in Pakistan.

The most important support of the party in the rural areas came from the middle or surplus farmers: they were in command of local party leadership and local government institutions and constituted a majority of the members of the national Parliament. Many of them also had trading and business interests, and in this respect they were linked up with the urban middle class.[1] Increasingly the party gained support in the upper echelon of the urban elite, the intellectuals and professionals; they became the strongest spokesmen in the fight within Pakistan for greater participation in economic opportunities. In the urban areas the party drew considerable support from the labour unions employed in the large-scale industry. Large-scale industry was predominantly owned by the Pakistani entrepreneurs. Therefore, to organise the workers in such industry against the Pakistani owners was a part of the nationalist struggle. By supporting and organising workers in their fight for higher wages and better working conditions, the Awami League obtained a powerful ally which was very useful in organising mass meetings and in mobilising popular support. The Awami League's opposition to "Big Business" and its support of the workers in the latter's struggle against "Big Business" had contradictory elements. Firstly, if the opposition to "Big Business" was carried into the period following independence, it was likely to alienate the aspiring, nascent Bengali industrialists who were trying to step into the shoes of the departing foreign capitalists. Secondly, by supporting the sometimes exaggerated claims of the workers against the Pakistani industrialists reaping high profits, the League raised expectations which it might prove difficult to fulfil when on the morrow of independence, industries would be owned either by the Bangladesh government or Bangladesh nationals. In its various public pronouncements and election manifestos before independence, the party was tending to advocate a policy of workers' share in profits and in the management of large-scale industry. Thirdly, it had acquired a political strength in excess of its numbers because of its concentration in large towns and its ability by organised action to dislocate the economic and administrative infrastructure in urban areas.

The Awami League, because of the composition of various groups supporting it, represented what some like Professor Kalecki had called "the intermediate regime". As against a regime ruled by large-scale industrialists and landlords, it was a regime of lower middle classes, composed of rich peasants, small industrialists and middle-class professionals, etc. This was a class which derived part of its income from scarcity rents, in the sense of surplus over opportunity costs. This was the case with professionals and bureaucrats with skills in limited supply. This was also true of small-scale industrialists and traders who earned scarcity premia on import licences or high profit margins on scarce domestic goods; in addition, there were the sharecroppers and surplus farmers, deriving rents from tenants and subsidies from the state. In the context of a poor country like Bangladesh, they constituted a privileged group and performed the role of a ruling class.

In the struggle for national leadership, the Awami League had succeeded in drawing a large following and considerable support from all sections of the population, including divergent economic and political groups. Since the objectives and programmes of the party were enunciated in general terms, it could accommodate political groups of both moderate "right" and moderate "left" persuasions. Among its members, there was a wide spectrum of political and economic philosophies. They were all bound together by one motivation, i.e. freedom from the political and economic domination of Pakistan.

However, with the passage of time, the radical elements among the workers and students started to exert pressure on the party to accept more radical and socialist policies and programmes. This was partly facilitated by the fact that during the late 1960s as the nationalist movement gathered momentum, students and workers became increasingly a significant force in the struggle for autonomy. During the period between 1958 and 1971, when Pakistan was directly or indirectly ruled by the military, political parties were either banned for a time during this period or even urban political parties were revived; the top leadership of the Awami League, the proponents of the autonomy demand, were in prison for most of the time with the result that the movement for autonomy for former East Pakistan and for the restoration of democracy had to be largely carried forward by the student activists. It was during the same period that the radical elements gained influence in both the student and labour wings of the Awami League. By 1968–69 a unity was forged between the Awami League student organisation (Student League) and the left-wing student organisations (Student Union, linked to the Moscow-oriented National Awami Party), to carry forward the movement for autonomy and democracy on the basis of a so-called "eleven point" programme, which absorbed the "six point" autonomy demand and included such measures as (*a*) nationalisation of banks, insurance companies and large industrial units, (*b*) reduction of taxes on agriculturalists, (*c*) payment of proper wages to labourers, (*d*) introduction of a flood control plan in East Bengal, etc.[2] At the same time, the trade unions organised by the Awami League extended their membership to include more radical sections of the working class movement. The workers, who were organised by the radical leftist parties, joined the Awami League-sponsored labour unions and constituted a pressure group for the adoption of radical policies to which the Awami League leadership had to respond. The war of liberation during 1970 was the last important factor in increasing the strength of radical elements in Bangladesh politics. The "guerilla" forces were drawn from young men of all political affiliations; young peasants and students, both rural and urban, were a very important component of the guerilla forces. There was a large contingent of radical elements among the freedom fighters, especially among the students. In the post-independence period, the leftist parties had gained in strength, especially with the acquisition of arms during the liberation struggle.

(b) Evolution of Thinking on Economic Systems and Policies

It is against the background of the above analysis of the evolution of the economic philosophy of the Awami League and the relative strength of the various interest groups supporting or opposing it that the attitude of the leadership towards economic development in general and the choice of an economic system and policies in particular can be understood.

The attitude of the ruling party to the choice of economic systems and policies in particular, whether to adopt socialism or a mixed economy, can be gauged partly from its election manifesto in 1970, the last election before independence, even though the main issue in the election campaign was not the choice of economic system but the demand for autonomy of East Pakistan.

Economic development problems and policies occupied an important place in the election manifestos and policy announcements of the ruling party before and after independence. It was the economic exploitation by Pakistan which swelled the tide of nationalism in East Pakistan. Therefore, the Awami League's popularity and appeal to the people of Bangladesh, rested on its promise of a higher rate of development and expansion of economic opportunities. An independent Bangladesh held out the promise not only to be able to conserve her resources, without their being drained away to Pakistan, but also to be free to choose policies and institutions to suit her own priorities, objectives and patterns of development. This was not possible under the colonial regime of Pakistan. The Awami League election manifesto in 1970 made specific pledges in the sphere of economic policy as follows:[3]

(1) establishment of a socialist economy and, as a first step towards the achievement of this objective, nationalisation of banking and insurance companies, jute industry, cotton textile as well as other basic and heavy industries;
(2) implementation of land reform and change in the system of ownership of land with a view to distributing land among the landless labourers;
(3) reform of the tax and revenue system and, specifically, abolition of the salt tax and land revenue in all landholdings below eight acres.

The next important step in the enunciation of the fundamental state policies in the economic sphere was the framing of the constitution for an independent Bangladesh in 1972. The constitution reflected the economic objectives of Bangladesh, as they were conceived by the ruling Awami party which framed the constitution then. The fundamental principles of the constitution were Democracy, Nationalism, Secularism and Socialism. The struggle first for autonomy, and then for independence from Pakistan was based upon the democratic principle of self-government and majority rule. The party throughout its struggle in Pakistan had consistently proclaimed its adherence to the principles of representative government.[4] Nationalism was the very basis of the creation of the independent state of Bangladesh. Because of the historical role of religion in the political development of the sub-continent, which was one of conflict and tension between the two major religious communities, and because twenty-five years of united Pakistan, which called itself an Islamic Republic, were marked by constant hostility between the two countries, India and Pakistan, the ruling party decided that there was no substantive advantage in calling itself an Islamic state. The examples of other states like Turkey and Indonesia, predominantly Muslim countries, which did not declare themselves as Islamic states served to lend support to such a policy. The economic implications of secularism were mainly a legal guarantee that because of religion no non-Muslim could be subjected to any discrimination.

More than any other principle of state policy, it was the principle of socialism which had important ramifications in the economic field. The constitution described the essential features of a socialist society as follows:

(1) an exploitation-free society and emancipation of the toiling masses from all forms of exploitation;
(2) the right to work on the part of every citizen and payment for work on the basis of the principle of "from each according to his abilities to each according to his work";
(3) equal opportunity for all citizens. "The state shall adopt effective measures to remove social and economic inequality between man and man and to ensure the equitable distribution of wealth amongst citizens and of opportunities in order to attain a uniform level of economic development throughout the Republic";
(4) discouragement of unearned income;
(5) the recognition of the basic responsibility of the State to obtain food, clothing, shelter, education, health care, jobs and social security for its citizens through planned growth;
(6) recognition of all laws expressly declared by Parliament to implement this basic responsibility, as constitutional in spite of their conflict with the fundamental rights listed in the constitution;
(7) limitation of private ownership of the means of production by law;
(8) inability of the Court of Law to challenge the nationalisation or requisition of private property;
(9) parliamentary democracy to facilitate the transition to socialism.[5]

Many of the propositions laid down in the constitution and stated above, especially principles 1–6, were common to almost all the liberal constitutions of a social-democratic type in the post-war world; most governments, irrespective of the type of constitution, seemed to accept or proclaim these principles. Principles 7–9 were more specific to a socialist state and referred to the specific features which were conventionally associated with the socialist state. An extension of the public ownership of the means of production was assigned a principal role in the transition of Bangladesh into a socialist state. Therefore, it did not provide for any fundamental rights in private property. The limits to and nature of private ownership would be determined by law, i.e. by the legislative process. The constitution provided three types of property, as follows:

(*a*) "State ownership, that is ownership by the State on behalf of the people through the creation of an efficient and dynamic nationalised public sector embracing the key sectors of the economy;
(*b*) Cooperative ownership, that is ownership by cooperatives on behalf of their members within such limits as prescribed by law; and
(*c*) Private ownership, that is ownership by individuals within such limits as may be prescribed by law".

There was no indication as to whether private and cooperative forms of ownership would be allowed only in the transitional stage or whether they would be in time completely overtaken by state ownership, excepting the very limited extent to which private ownership, without wage employment, was allowed to function in a few socialist countries. Legally and logically the provisions of the constitution were consistent with the abolition of private ownership altogether. Since the limits to private ownership were to be set by law, there was no restriction on the Parliament to pass a law which would set the limit at zero. The evolution of the relative importance of different forms of ownership was left to the parlamentary process and to the decisions of the majority in the Parliament.[6] There were those in the government and the ruling party who believed that the co-existence of private, co-operative and state enterprises should be a permanent feature of the economy. At the same time, others believed that the co-existence of the three forms of ownership was a transitional phase until the goal of socialism was reached, and that under socialism, there would be only state enterprises and workers' co-operatives and no wage employment, i.e. no exploitation of labour by private capital.[7] Moreover, there was no consensus on the role of private property in the transitional stage. The majority in the ruling party tended to veer towards the former, i.e. co-existence rather than the latter position.

The role of private property, however, would depend on whether it was to be a long-term feature of the economy or a transitional phenomenon. If the ultimate goal were the establishment of socialism, strengthening of private enterprise in a transitional phase could create contradictions which could only be contained if there were unanimity or a consensus as to the goal and a determined leadership committed to it, so that the role of private enterprise could be reversed whenever necessary. There was no consensus on the role of the state and private enterprise, once the transitional phase was over; therefore, those who believed in the ultimate elimination or the very limited role of private enterprise in the long run, feared that a significant role for private enterprise in the initial stage of economic and political development of Bangladesh would only result in strengthening forces against socialism. There was, therefore, a very difficult choice and a dilemma for those who somehow had faith in the evolutionary path to the establishment of socialism in the long run. There were those among them who considered the extension of the public sector as beyond the administrative and managerial capacity of the government machinery and who regarded suppression or underutilisation of the entrepreneurial capacity of the private sector as inimical to the possibilities of overall growth. But at the same time there was no agreement as to whether Bangladesh really had a substantial reservoir of untapped private enterprise and initiative for the development of large-scale industrial and commercial activities. Lack of opportunities during the Pakistani regime had stunted growth possibilities. Whatever private enterprise was potentially available in Bangladesh, it was argued by many, could be used to best advantage in the small-scale industrial sector and in the agriculture-related activities.

There was also uncertainty relating to a provision of the constitution which stated that unearned income would be discouraged. There was no clear definition of what was considered "unearned income". Was all income from property unearned income? Was interest unearned income? Were rents earned by landlords either from sharecroppers or from tenants unearned income? Since private enterprise was to play a positive role, profits from ownership of capital and entrepreneurial ability could not obviously be considered unearned income. The windfall profits earned from the sale of import licences could be genuinely called "unearned income".[8] The constitution said nothing about feudalism, income from land, sharecropping or absentee landlordism; nor did it pronounce its ideological position *vis-à-vis* the role of foreign private capital. These two aspects of economic policy have often been important components of the socialist platform in various other developing countries.

The lack of consensus as to the nature of the socialist economy which Bangladesh should establish in the transitional phase as well as in the long run was thus not a matter of differences between the ruling party and the opposition parties; there existed differences of opinion between the factions of the ruling party and among the members of Parliament as well as of the cabinet. The leftist opposition parties did not have faith in the evolutionary path to socialism. The leftist criticism of evolutionary socialism, through prolonged co-existence of the private and public sectors, was that so long as the ideological foundations were not changed, and the profit motive remained the prime mover in a large part of the economy, nationalisation (to the extent that it was carried out) would be used to promote the interests of the private sector because the state apparatus itself was not in the hands of the poor and property-less. This was the belief of almost all leftist parties except a few such as the Pro-Moscow National Awami Party. In the case of the latter, it was not clear whether their support of the Awami League's pronounced policy of following an evolutionary path to socialism was a matter of tactics or of faith. They could have taken the view that in the absence of strong revolutionary forces working towards socialism in Bangladesh, the best way was to ensure that the Awami League would pre-empt the growth of large-scale capitalism in Bangladesh, by developing a large public sector and by preventing the penetration of powerful multinationals in the private sector of the economy.

Questions were raised as to whether democracy was consistent with socialism in Bangladesh. So long as socialism was to be achieved through an evolutionary process, by popular will and consensus, democratic institutions were a necessity for the emergence of socialism in Bangladesh. They would allow the various parties and groups with a belief in socialism to propagate socialism through the exercise of democratic rights of freedom of speech and association; it would enable them to persuade others by peaceful means through the laws passed by the Parliament. If there were no democracy and if political power was to be acquired through force, it could very easily fall into the hands of the non-socialist fascist groups, in the absence of revolutionary socialist parties. If the Awami League were to establish socialism by force and establish a one-party state, the party itself could split, because there was no firm conviction or concensus within the party on the nature, time path and the manner of establishment of socialism. The achievement of socialism by democratic means and by a gradualist approach was not without the possibility of contradictions. In a democracy where both the socialist and non-socialist parties freely operate, the socialist policies, such as for example, nationalisation of and restrictions on the operations of private enterprises, undertaken by a socialist party in power could be reversed by the next party in power if it happened to be non-socialist. This would not only dislocate productive activity, by for example, the transfer of productive enterprises between public and private ownership; it would also create uncertainty about the future economic policies of the government and discourage decisions to carry out investments with longer gestation lags. In the absence of a general consensus among all the political parties about the goal of socialism, there was the possibility that the constitution itself would be amended and the fundamental rights to private property restored, while the right of the Parliament to put limits on private enterprise could be abrogated. Because of the flexibility of the socialist provisions in the constitution and because of the democratic institutions there were serious risks of wide swings in the choice of economic institutions and policies, as the pendulum swung from side to side.[9]

The manifesto of the Awami League during the elections in 1973 under the new constitution laid down more detailed guidelines and policy directives. The main economic policy issues enunciated by the election manifesto were as follows:

(1) Development under a regime of economic planning was an essential prerequisite for the advancement of Bangladesh. This was also an integral part of of socialist development.
(2) A number of actions which had already been taken constituted only the first stage in the progress towards a socialist economy, which would be accelerated under the Five-Year Plan with a view to expanding people's ownership of the basic industries and other forms of productive activity. These actions included:

(*a*) Nationalisation of banking and insurance companies. The jute, cotton textile and sugar industries had already been nationalised on 26th March 1972. In addition, several industries such as shipbuilding, heavy iron and steel products, cement, paper and newsprint, etc., were already under public ownership. These consisted of enterprises abandoned by Pakistanis and put under public ownership after independence.
(*b*) The decision to place under public ownership the great bulk of the abandoned industries as well as foreign trade.
(*c*) Imposition of a ceiling on private investment at Tk.2.5 million and restriction of foreign private investment to the public sector with a view to expanding investment in the public sector.
(*d*) Establishment of "fair price" shops in every union (clusters of 10 villages with a population of 10,000 to 12,000) with a view to securing balanced and equitable distribution of essential consumer goods to the people at large.
(*e*) The decision to formulate laws and regulations which would enable some participation of the labour force in the management of the nationalised industries, so that it would consider itself a partner in the productive system of the country.

The election manifesto reiterated the party's programme of evolutionary socialism and its faith in economic planning. In addition to the steps already taken for the expansion of the role of public ownership, it promised (*a*) a greater role for the public sector in the marketing of essential consumer goods and (*b*) the participation of labour in the management of public sector industries. Even before independence, the Awami League had talked about the participation of labour in the management and profits of the large-scale industrial sector. Before independence, its advocacy for the participation of labour in profits and management related to both public and private sector industries; in fact, at that time the private sector large-scale industry was owned predominantly by the Pakistanis and worker-participation in profits would have given East Pakistan a larger share of income generated in her industries and would have reduced the drain to West Pakistan. The shift in announcing policies for the public rather than private sector industry only could partly be explained by the fact that all large-scale industries were already in the public sector and within the ceilings imposed on private investment, private enterprise was expected to operate only in the small-scale sector.

(c) The Five-Year Plan and Socio-Political Perspectives

It was not only the constitution of Bangladesh and the manifesto of the ruling party that attempted to set out the guidelines relating to the future economic system of Bangladesh; the Five-Year Plan prepared and published in 1973 also attempted to elaborate the socio-political perspectives at length. In fact, the Plan attempted to set forth the socio-political aspects of Bangladesh's future course of economic development on the basis of its understanding of the public pronouncements of the political party in power. It elaborated the implications of a policy of transition towards socialism which the political leadership accepted as the constitutional goal. As already mentioned, a considerable enlargement of public ownership of the productive sectors in the economy already took place in 1972, before the formulation of the Plan in 1973. The Plan suggested measures which would carry further forward the consolidation and enlargement of the public and co-operative sectors. The most important areas of private enterprise were (*a*) trading, including both internal and foreign trade, and (*b*) agriculture. The Plan suggested that lower ceilings on landholdings should be seriously considered in the future, and meanwhile, attempts should be made to strengthen the economic position and bargaining power of the poorer farmers and the landless labourers. The rural co-operatives were to include both the small farmers and landless labourers; the latter was to be accorded a voice in the management of the co-operatives, in proportion to their numerical strength in the rural society.[10] Furthermore, the Plan suggested the extension of co-operative enterprises in various fields, especially internal distribution, housing and construction; it advocated tax measures to siphon off excess profits or scarcity premia on imported or domestically produced goods in short supply and taxes on wealth and income were recommended not only to mobilise resources for public investment but also to reduce the inequality of income from work and property. The Plan accepted a continuing "mixed economy" as a transitional stage in the evolution of a socialist society in Bangladesh. It postulated that changes in institutions and in economic policies should be introduced in such a manner as to avoid serious or sudden dislocation in the economic system or in the tempo of economic activity. The prevailing system was to be changed in a way and at a pace which would enable Bangladesh to develop new attitudes and policies and to manage new institutions with reasonable efficiency. In the near future and particularly during the period of the Five-Year Plan, the objective was to achieve a more equitable distribution of income, partly through redirecting public expenditures towards the poorest sections of the population and partly through increasing the role of co-operatives and the public sector in trade, industry and services. The need to use material incentives to stimulate initiative and enterprise and to ensure efficiency of productive activities was emphasised. The Plan stressed the role of prices as a signal and incentive mechanism for securing a rational allocation and an efficient utilisation of resources. Exceptional effort and scarce skills would require financial incentives, and in the interest of increased production such rewards or incentives were recommended as a part of conscious state policy.

The ruling party, it was suggested by the Plan, should set about organising cadres, i.e. young dedicated workers who believed in an evolutionary transition to socialism and were willing to participate in development activities and to identify themselves with the masses at the grassroots level; they were to mobilise, inspire and organise the masses for productive work and to prepare them for accepting changes in income distribution, property relationships and new forms of economic organisations in agriculture, trade and industry. The role of cadre in a multi-party parliamentary democracy was not an unambiguous one. The Plan rejected the idea that the established bureaucracy or civil servants could serve as the principal agents for social transformation; the latter by virtue of their training, habits of work and methods of approach were on the side of status quo and established traditions—innovations and changes were not their forte. Moreover, there was the need to mobilise the masses as well as to ensure their effective participation in rural development projects; the trade unions were to play a constructive part in increasing production, while at the same time gaining a greater participation in the decision-making process. In the fully fledged socialist countries, the cadres were incorporated in the administrative and decision-making machinery as an integral and internal component. In Bangladesh, they were to be considered as the watchdog or catalytic agents rather than as part of the executive agencies; they would basically remain as outsiders. The concept of a group of committed party workers or cadres and that of a neutral bureaucracy, which would function without any party affiliation or convictions, as implied in a multi-party parliamentary democracy, were not wholly consistent. They were to function side by side. The former was to stimulate, goad and supervise the latter and to ensure that the latter faithfully carried out the policies determined by the political leadership. This was a novel experiment fraught with uncertainties.[11]

There was an implicit assumption throughout that the top leadership of the ruling party, particularly the Prime Minister because of his great popularity with the masses, his command over the party structure, and his towering and charismatic personality, would be able to hold an appropriate balance between the divergent interest groups in the country and the various factions in the party. He would be able, wherever necessary, to resolve the conflicts or contradictions with a view to steering the party and the country along the evolutionary path towards socialism. A few interest groups like the students and labour unions were organised politically; a few others such as the army and the bureaucracy were organised institutionally and bound together by a broad range of common attitudes and interests, while others like the surplus farmers or trading classes were not organised either politically or institutionally but had close links with the different groups within the party as well as within bureaucracy and the army. The armed forces in Bangladesh, especially those who participated in the war of independence, were politically orientated; they had worked in close collaboration with the political groups of divergent persuasions and convictions during the war of liberation and those links were not completely terminated in the post-independence period.

The Plan recognised that in view of the lack of consensus within the ruling party about the evolutionary time path to be followed and the final content of the socialism which was pronounced by the party as its goal and which was outlined in the constitution, the evolution of the socio-economic system would be determined largely by the extent to which the Prime Minister was able to resolve the differences or conficts between the interest groups and ideologies within the party, and by the extent to which he would be willing and able to throw his personal authorty and weight on the side of one group or another. The outcome of the resolution of intergroup conflicts would partly depend upon the conviction of the party leadership, especially the Prime Minister, and partly upon the relative strength and popular following of the divergent interest groups. The predominant consideration in this balancing act would be to ensure the survival of the party and its continued leadership. The Plan laid down the socio-political perspectives of development on the implicit assumption that the top leadership of the party, especially the Prime Minister, would not succumb to the temptations of an easy option when faced with the hard choices that had to be made for mobilising and using them productively for development in an egalitarian and socialistic framework. It was hoped that the top leadership of the party would practise self-sacrifice and rigid discipline as well as demonstrate its faith by positive action in conformity with what it preached. It was recognised that if the leadership was weak and vacillating, it would have a difficult task in enlisting constructive co-operation from the people or from the rank and file of the party who lacked very strong ideological commitment in any case, in the uphill task of economic development in a socialist framework.

(d) Political Commitment to Economic Development and Planning

Enough has been said in the foregoing to indicate that economic development constituted an important plank of the platform of the Awami League before and after independence. The historical circumstances of the creation of the independent state of Bangladesh, which drew its main inspiration from the economic exploitation of Bangladesh by Pakistan, could not but result in the party attaching great importance to the objective of economic development.

The election manifesto of the party in 1973 made the most comprehensive and detailed statement on the economic problems and policies of Bangladesh. The pronouncements in the election manifesto covered all sectors of the economy; among the important development objectives reflected in the manifesto were the following:

(1) Acceleration of the pace of the green revolution in view of the critical importance of agriculture in the Bangladesh economy; assured and adequate supplies of agricultural inputs such as fertilizer, pesticide, implements, etc., and emphasis on domestic production in the public sector of such agricultural inputs with a view to obtaining security of supplies; prevention of the drain on domestic resources consequent on a large and growing volume of food imports, by the achievement of self-sufficiency in food production in Bangladesh;

(2) Establishment of a separate jute ministry in order to promote the development of all aspects of the jute industry and trade including

development of research facilities; nationalisation of foreign trade in raw jute, and public ownership of the major part of the internal marketing of raw jute with a view to ensuring a fair price to the growers;
(3) Improvement in the efficiency of production of important cash crops such as tea, sugar and tobacco; further development of the manufacturing industries based upon these three raw materials;
(4) Development of fishing, a means of livelihood for an important segment of the rural labour force, through the (*a*) development of a fleet of trawlers for deep sea fishing, (*b*) modernisation of fishing equipment, (*c*) development of inland fisheries, and (*d*) the supply of yarn for the fishing nets and of materials for fishing boats for this purpose;
(5) Flood control and extension of irrigation facilities;
(6) The ceiling on landholdings, already reduced from 125 acres to 33 acres in 1972, would be further reduced only as improvement in agricultural productivity took place over time; abolition of private leaseholdings of the rural markets (hats and bazaars);
(7) Rural development with particular emphasis on (*a*) the rural works programme, (*b*) the organisation of co-operatives for handloom weavers and the direct supply of yarn to weavers without the intermediary of middlemen, (*c*) the establishment of banking facilities in every Thana, (*d*) tubewells for the supply of pure drinking water (one tubewell for every thousand people), (*e*) the development of ideal villages—one in every Thana, and (*f*) the excavation of rural tanks (one in every union) and the development of farming operations on the banks surrounding the tanks;
(8) In the field of industrialisation, achievement of self-sufficiency in the domestic production of the following items: edible oil, sugar, kerosene and fuels, cotton cloth, fertilizer, medicine, cement, building materials, iron and steel products, agricultural implements, petroleum and certain petrochemicals, gas, etc,;
(9) Establishment of the Cottage Industries Corporation with a view to encouraging handloom weaving as well as other indigenous cottage industries and development of cottage industries through co-operatives;
(10) Development of transport by road, waterways and railways with special emphasis on (*a*) a bridge over Jamuna, linking the northern and southern parts of Bangladesh, (*b*) domestic production of trucks and buses, (*c*) imports and domestic production of marine engines for inland water transport, (*d*) expansion of Chittagong port and development of a fully fledged port in Chalna and (*e*) dredging of dried up waterways for the improvement of inland water navigation;
(11) Rural electrification for puposes of agricultural and industrial development;
(12) Acceleration of family planning programmes for the control of population; creation of employment opportunities and formulation of development programmes with a view to maximising employment opportunities;
(13) Extension of educational facilities with emphasis on the development of manpower and vocational and technical training facilities for meeting the requirements of economic growth; establishment of an Education Commission consisting of experts to lay down the guidelines of and priorities for the educational development of Bangladesh;

(14) Development of rural health centres, new medical colleges and nursing schools with a view to extending (*a*) cheap medical and health facilities to the entire population and (*b*) free facilities to the majority of the population within the next five years.[12]

As is obvious from the above, the pronouncements of the election manifesto consisted of many proposals which were very general in nature and therefore, were not in the nature of pledges, for which accountability could be made specific or clear. However, there were other proposals which were in the nature of detailed programmes and policies, including specific projects, ranging over the various sectors of the economy. It was not that the party or its leadership had undertaken a close examination of the individual programmes and projects. Many of these proposals, however, had been discussed in a general way for some time. More importantly, while the manifesto was being written, the Five-Year Plan was under preparation. Most of the proposals were under consideration in the Planning Commission and a few of them were discussed with the ministers and the Prime Minister. They appeared generally attractive and reasonable propositions. Accordingly, they were incorporated in the manifesto. The proposals about cottage industry, rural electrification, food self-sufficiency, population control, etc., were general in nature and had been the components of the generally accepted development strategy even in the pre-independence days. The specific proposals in these sectors which were at the time under discussion between the Planning Commission and the relevant ministries were incorporated in the election manifesto.

The fact, however, that so much—in fact the major part—of the manifesto concerned itself with economic policies and programmes, was undoubtedly a reflection of the importance in a general sort of way which the political leadership accorded to the economic development of Bangladesh, even though they had not seriously thought out the alternative strategies or the institutions and policies, which would be required for the accomplishment of the objectives. The policies and programmes had economic as well as socio-political costs and benefits in terms of their impact on different groups and classes in society. especially the interest groups on whose support the party depended for its political strength and, indeed, survival. The relationship between the declared goals of economic development and concrete policies and institutions was very vaguely perceived. Moreover, specific commitment on concrete policies and institutions was frequently not called for either when the election manifesto was announced or while the Five-Year Plan was being prepared. It was only when actual decisions had to be taken from day to day that the political leadership was confronted with the task and the challenge of formulating and taking concrete decisions on policies and institutional changes. It was at this stage that the costs and benefits of alternative policies and courses of action, and particularly the impact on different interest groups, were brought to light and that the political leadership was called upon to respond to the demands of conflicting interests.

When appraising or evaluating the commitment of the political leadership towards economic development or a socialist economy, it is important to recognise that economic objectives were not and could never be the only objectives of a party or a government in power. There were non-economic goals which the government had to pursue, some of which were very important in the early years of independent Bangladesh. The pursuit of these goals, by the nature of the case, was not, and could not always have been rendered subservient to the overriding objectives of economic development. These goals demanded and deserved considerable time and attention from the political leadership, in competition with economic objectives. Moreover, some of them were prerequisite for the achievement of economic development. In this category one could consider the objective of maintenance of law and order and the preservation of the security of life and property, which broke down during the war of liberation. A large number of people and interest groups of various political affiliations had acquired arms during the war of liberation. There were also convicts and criminals who escaped from prisons during the Pakistani regime, when prisons were thrown open by the political activists as an act of reprisal against the occupation army, and who had acquired arms. In spite of successive attempts on the part of the government to recover them, large amounts of unauthorised arms were left in the hands of various political, as well as non-political, groups and individuals. Law and order was in serious jeopardy. The principal task of the government in the period following liberation was to establish law and order and to strengthen the law-enforcing agencies, i.e. the police and the para-military forces, which were disorganised and largely disintegrated during the war of liberation.

Another related problem was that of "collaborators", i.e. individuals and groups who opposed the creation of independent Bangladesh and who collaborated with the Pakistani regime during the war of liberation. While some of them were to be tried and punished, a large majority of them were to be integrated in the socio-economic life of Bangladesh. There were those among the "collaborators" who had specialised skills and constituted an important segment of the skilled labour force; there were a few others who had provided the bulk of the managerial or supervisory personnel in a few sectors of the economy in pre-independence Bangladesh. This was particularly true of a large number of the non-Bengali speaking "Biharis" who had migrated from India to Bangladesh (then East Pakistan) at the time of creation of Pakistan in 1947. Many of them were ideologically opposed to the creation of an independent state of Bangladesh. Repatriation to Pakistan of those among them who wanted to settle there was a subject of negotiation with Pakistan and the outcome was highly uncertain. In the meantime, they had to be housed and fed in camps, where in the aftermath of the liberation of Bangladesh, they had been placed after being uprooted from their homes. To the extent that they could not be repatriated to Pakistan, they had to be re-absorbed in the socio-economic life of Bangladesh.

There were two more important political tasks facing the political leadership. One was to frame a constitution for the new country and the other was to hold general elections under the new constitution. A great deal of time and effort was devoted to the making of the constitution. There were public debates and discussions on the principles and articles of the constitution, not only in the constituent assembly, but also in various public fora and news media. As far as the need for a new election was concerned, it was partly due to the fact that the context of the earlier election was very different. The achievement of autonomy for East Pakistan within the context of Pakistan was the platform on which the election was fought in 1970. In the new state, issues relating to internal and external policies, institutions and relationships needed reappraisal and either alteration or confirmation. A fresh mandate from the people was expected to provide a boost to the confidence of the party in terms of popular support for its policies and programmes. Already there were complaints about corruption and nepotism in the party and about its possible interference in the day-to-day administration of the country. A vindication for the party through popular support at the elections was also considered to be politically advantageous.

Apart from the above non-economic objectives and priorities engaging the attention of the political leadership on the domestic front, there were issues of similar importance in the field of foreign policy. Bangladesh as a new state had to secure recognition from as many nations as possible; she started her independent existence with diplomatic recognition from only a very few states. The various governments of the world, especially the big powers, were divided in their attitudes towards the creation of Bangladesh and had played divergent roles during the war of liberation. Bangladesh had to establish in relation to all of them her identity and credibility as an independent state. Furthermore, the relationship with Pakistan had to be normalised in the interest of peace in the sub-continent. The Pakistani prisoners of war had to be sent back and Bengalis from Pakistan had to be repatriated. The settlement of financial assets and liabilities of the previous Pakistan State (East and West) were important unresolved issues even up to the middle of 1975. The above is only a selection from a long list of non-economic tasks and goals demanding the government's urgent attention in the field of both domestic and foreign policy objectives. A few of them had important interrelationships with economic development, but many others which did not nonetheless deserved high priority.

Notes

1. During the 1970 elections, 72% of the nominees of the Awami League for the National Assembly elections consisted of lawyers (47%), doctors (4%), businessmen (19%) and retired government servants (2%). T. Maniruzzaman, *Radical Politics and Emergence of Bangladesh*, Bangladesh Books International Ltd., Dacca, 1975, p. 40. The rural government institutions by the late 1960s came to be dominated by affluent or surplus farmers. It is they who joined the Awami League in large numbers during the movement against Pakistan, in the late 1960s. 61% of the Chairmen of the Union Councils (lowest local government organisation) in 1970 consisted of those with an income of Tk.4,000 and above, and 76% consisted of those with incomes above Tk.3,000. A. Abdullah, "Land Reform and Agrarian Change in Bangladesh", *Bangladesh Development Studies*, January 1976, p. 91.

2. T. Maniruzzaman, *Radical Politics and the Emergence of Bangladesh*, op. cit., p. 37.

3. A. Mannan, *Manifesto of All Pakistan Awami League*, Publicity Secretary, All Pakistan Awami League, 1970.

4. In the light of subsequent developments in early 1975, when one-party government was established by its leader, it was clear that its commitment to democratic principles and institutions was not without reservations or at least that, whatever their original aspirations, in the years following independence there was a change in the attitude of its leadership.

5. "The constitution provides the mechanism by which Parliament, democratically elected by a politically conscious people, can determine priorities on the basis of objective conditions obtaining in the country at a particular phase, and implement them consistently with socialism and the wishes of the people." T. Maniruzzaman, "Bangladesh: An Unfinished Revolution", *Journal of Asian Studes*, August 1975, p. 893.

6. ibid. The constitution of Bangladesh reflected the balance of forces. Within limits, the choice of economic system, policies and instruments, was kept flexible, in view of the wide spectrum of political and economic philosophies embraced within the ruling party. It had contradictory forces pulling in different directions. Moreover, it was at the same time facing an increasing threat from the leftist parties who had command over arms and who did not believe in the evolutionary path to socialistic transformation.

7. To start with there was uncertainty and confusion as to the role of co-operatives in a socialist state in the transitional stage as well as in the

final stage. There were those who argued that any enterprise in any field of activity, without limit of size in terms of capital and employment provided it was run on the co-operative principles, was consistent with socialism. The co-operative principle was said to consist of the following: (*a*) one member with one vote, irrespective of the relative share held by a member in the equity capital and (*b*) a minimum of fourteen members, as required by the prevailing co-operative law. There were, on the other hand. those who argued that a co-operative of this nature was barely distinguishable from a joint-stock enterprise and private capitalism, excepting that it was more democratic, i.e. voting rights were not related to the numbers of shares held by members. A joint–stock company with equal shareholding for all members was exactly the same as a co-operative and, in this view, this was not consistent with the spirit of a constitution which professed to make a transition to socialism. Moreover, co-operatives were not to be allowed to employ wage labour, if at all, beyond a certain maximum limit.

8. Because of exchange control and excess demand for scarce imports, the domestic prices of imports exceeded the duty paid landed cost of imports. In view of the lack of enthusiasm of the ruling party against the taxation of land rents or scarcity premia on import licences, it was doubtful whether the provision of the constitution was understood clearly and if understood, whether it was taken seriously.

9. The constitution did, however, try to make such fundamental swings in policy difficult. "No bill which contains provision for, or has the effect of, divesting the State of any property or of enhancing any compensations payable by the State shall be presented to the President for assent unless it is passed by votes of not less than two-thirds of the total number of members of Parliament". A. F. Huque, "Constitution-Making in Bangladesh", *Pacific Affairs*, Vol. I, 1973–74.

10. "It is well known that serious class differences exist in the rural society of Bangladesh. The rural power structure is authoritarian, dominated by the vested interest groups. If these groups continue to dominate and make decisions regarding choice of new technology and institutions, the interest of the lower stratum and the majority of the rural society will not be safeguarded, class differences may be intensified, and bitter class struggles will become inevitable. This problem can only be solved by helping the depressed classes to organise themselves. adopt innovations collectively and become a dominant collective force". *Five-Year Plan*, p. 156.

11. As subsequent events unfolded, Sheikh Mujib by 1975 abandoned the concept of multi-party parliamentary democracy and adopted the policy of a one-party state, in which the party cadres and bureaucracy, including the army, were to be integrated under one unified leadership and control.

12. *Manifesto of the Bangladesh Awami League—1973* by Mr. Zillur Rahman, General Secretary, 19 February 1973.

2. Planning Machinery in Bangladesh

(a) Genesis of the Planning Commission

Planning had a long history in pre-independence Bangladesh. The Pakistan Planning Commission came to occupy an important place in economic decision-making in Pakistan. Its status and importance was indicated by the fact that the President of Pakistan himself was the Chairman of the Planning Commission. The leadership in Bangladesh remembered that the Pakistani Planning Commission had played an important advisory role in deciding the allocation of investment resources between East and West Pakistan. In the mid-1950s the Awami League had been briefly in power in the provincial government of East Pakistan. It was then that a Provincial Planning Board was established. It was seen as an important instrument for formulating investment programmes and for negotiating with the Central Government of Pakistan for an adequate share of the financial resources available to be used for the development of East Pakistan. The Board undertook the task of appraising and evaluating East Pakistan development projects which fell below a certain size.

Thus, in the light of its historical experience, the political leadership was generally aware of the need to establish a planning machinery in Bangladesh. Furthermore, during the struggle for independence the political leadership of the Awami League became closely associated with a number of economists and sought their advice in formulating and articulating their programmes and objectives, especially during the 1960s. It was thus to be expected that on the creation of independent Bangladesh, the political leadership would establish planning machinery and that it would turn to those close economic advisers associated with it during the struggle for independence, to man it. In fact, the Planning Commission which was established in January 1972, had its embryonic start, at least conceptually, if not in the full operational sense, in the Planning Cell which was established in the Mujibnagar government, that is the government-in-exile which was established in India during the war of independence. The Planning Cell established during the war of liberation was mainly concerned with formulating a programme, both physical and financial, of reconstruction and rehabilitation of the economy of post-independence Bangladesh.

(b) The Organisation, Nature and Functions of the Planning Commission

One of the first tasks of the Prime Minister, on his return from prison in Pakistan to head the new government of Bangladesh, was to establish the Planning Commission and to appoint the Deputy Chairman and members of the Commission, who were asked to work out the organisational structure and the functions of the Planning Commission. The cabinet decision establishing the Planning Commission stated the functions of the Commission as follows:

(1) To prepare national plans, annual, five-year and perspective, for the economic and social development of the country in accordance with the socio-economic objectives of the Government of the People's Republic of Bangladesh.
(2) To prepare annual development programmes within the framework of the national plan, and to formulate policies for the implementation of the plans.
(3) To evaluate plan performance and watch the progress of plan implementation on a continuous basis in order to prepare evaluation of national plans.
(4) To study important economic issues and formulate economic policies and measures.
(5) To determine external aid requirements and negotiate the total quantum and composition of aid required from foreign countries.
(6) To evaluate external debts and submit reports thereon along with evaluation of national plans.
(7) To undertake and promote economic research and to initiate surveys and investigations needed to support effective planning and development.
(8) To advise on the nature of the machinery for securing the efficient execution of the national plan, annual programmes, and economic policies.
(9) To stimulate and where necessary, initiate project formulation; to examine and advise on programmes and projects with a view to determining their consistency with the national objectives.
(10) To review the progress of implementation of approved projects, particularly of aided projects; to identify the causes of delays and difficulties in the implementation of projects and to prepare solutions thereof.[1]

It is apparent from the above that the Planning Commission sought to perform three broad sets of functions. One was to prepare the annual and longer term plans, including five-year and perspective plans; in this connection, it was to assist the ministries in preparing development projects. When large development projects covering several ministries were involved, the Planning Commission was to take the lead by undertaking feasibility studies and co-ordinating the preparation of projects jointly with a number of ministries. Moreover, as a part of its task of preparation of the development programme, it was to undertake technical and economic appraisal and evaluation of projects for the purposes of their inclusion in the Plans. Secondly, since a Plan was to consist not only of intra-sectoral and inter-sectoral financial allocations, but also of recommendations on a whole range of policies and institutional changes which were necessary for the implementation and realisation of the objectives of the Plan, the Commission was expected to make recommendations as well as being involved in the process of deliberation on these matters. Thirdly, the Planning Commission was to be a central agency for co-ordination of economic policies, both short- and long-term, which were to be undertaken by the various ministries, as for example, the determination of import, export, exchange rate and price policies. Moreover, in all policy matters the National Economic Council, consisting of the main economic ministers of the Cabinet and headed by the Prime Minister, was created to serve as the economic mini-Cabinet for deliberations on economic policy issues and for the approval of development projects and programmes. The Planning Commission was to serve as its secretariat for major economic policy questions and for initiating the appraisal of development projects and programmes by the National Economic Council.

The structure and organisation of the Planning Commission corresponded to the functions it was to perform. There were ten Divisions in the Planning Commission, each headed by a Chief enjoying the status of a Secretary to the government of Bangladesh, the highest ranking civil servant in the administrative hierarchy. These Divisions were: General Economics, Programming and Evaluation, Industry, Agriculture, Water Resources, Rural Institutions, Physical Infrastructure, Socio-Economic Infrastructure, External Resources and Administration. For example, the Programming and Evaluation Division was in charge of co-ordinating the task of appraisal of all development projects, before they were sent to the National Economic Council for approval. This task it performed in co-operation with the technical or sectoral divisions of the Planning Commission (such as Agriculture and Industry, etc.) and the administrative ministries and agencies in charge of formulation and implementation of development projects. The General Economics Division was concerned with all matters relating to economic policies, such as tax and expenditure policy, foreign trade and exchange policy and price policy, mobilisation of both domestic and foreign exchange resources for the Annual and Five-Year Plans, etc. This Division was also in charge of preparing the evaluation and appraisal report on the performance of the economy. The General Economics and Programming Division put together the sectoral projects and programmes in a comprehensive framework for the purposes of the preparation of the Annual and

Five-Year Plans. The General Economics Division was to provide the macro-economic framework for the preparation of the Annual and Five-Year Plans; it worked on an estimate of resources, both domestic and foreign, and on the size and sectoral priorities of the investment programme. The External Resources Division, as its named implied, was concerned with mobilising resources from abroad for the implementation of the Plan. It was the only Division of the Planning Commission which was executive or administrative in nature; it was not involved merely in making recommendations but also in actually carrying out detailed aid negotiations for projects and commodities as well as supervising and monitoring the progress of utilisation of foreign economic and technical assistance.

The other Divisions of the Planning Commission were more technical in nature. They were in charge of programmes and policies relating to the specific sectors of the economy; they were, so to speak, the counterparts of the various ministries dealing with particular sectors of the economy. The Industries Division of the Planning Commission, for example, was the counterpart of the Ministry of Industries and dealt with the policies and progammes relating to the public and private sector industries. The sectoral Divisions of the Planning Commission were to prepare sectoral development programmes and projects in collaboration with the ministries, to undertake the evaluation of projects as well as to report on the progress of implementation of development projects.

The Deputy Chairman and the Members of the Commission, who enjoyed the status respectively of Minister and State Ministers, co-ordinated and supervised the work of the various Divisions. Each Member, of whom there were three, was in charge of a number of related Divisions, for example, one Member was in charge of industry, physical infrastructure, power and natural resources; another Member was in charge of agriculture, rural institutions and water resources, etc

The Planning Commission was in a sense a macrocosm in a microcosm, containing the various categories of services and professions encompassed by the entire governmental machinery. Partly because it included civil servnts from various ministries and government agencies, the Planning Commission itself was not immune to inter-service rivalries and conflicts existing elsewhere in the administration. Because the General Economics Division was concerned with the analysis of general economic policy issues, it was manned exclusively by economists. Other economists were scattered throughout the sectoral or technical Divisions. The original idea of having economists in these Divisions was to involve them intimately with the preparation and evaluation of programmes and projects; they were also intended to impart to the specialists in the various technical Divisions, a familiarity with economic criteria and an appreciation of their role in the formulation of the sector programmes and projects. The technical specialists were, in turn, expected to impart their concepts and criteria to the economists who could hopefully acquire over time a detailed knowledge of the subject-matter. The technical Divisions were to become interdisciplinary teams with different specialisations mutually reinforcing each other. In practice, however, there were often tensions and strains between the economists and the "subject-matter" specialists who claimed a position of dominance and leadership. Moreover, in view of the dearth of qualified economists, there were many technical divisions without any economists. Serious thought was, therefore, given at a later stage to the suggestion that the economists should all be placed in a central pool located in the General Economics and Programming Divisions, so that the interaction and feedback between economists and technical specialists could be promoted through interdivisional consultation rather than intra-divisional collaborative work.

The Planning Commission suffered from a shortage of staff throughout the period under review. This was especially acute in the case of the social sectors. Moreover, in the technical Divisions there was a shortage of personnel to undertake the monitoring of the progress of implementation of projects. Among the functions of the Planning Commission, as originally conceived, this was an important aspect of its responsibilities in order to secure adequate feedback between the problems of implementation in the field and the formulation of an investment programme. This feedback, as it happened, was obtained mainly through the reports of the ministries rather than through independent enquiries and investigations of the Planning Commission itself. In view of the acute shortage of qualified personnel, many technical experts, such as engineers, scientists, doctors were drawn temporarily on secondment from the executive of administrative ministries. Since they had actual operational experience of the various sectors, their knowledge and experience was expected not only to enhance the efficiency of the Planning Commission but also to promote a close working relationship with the administrative ministries. This arrangement, however, had one limitation. There were instances when specialists became partisans for the ministries from which they came. Their neutral role was compromised by the fact that many of them were only temporarily assigned to the Planning Commission and were expected to return to their original ministries and agencies. If they were to function conscientiously and with strict objectivity in their scrutiny and evaluation of the projects and proposals of the ministers, they were afraid that they might annoy and, therefore, damage their own future prospects in their respective ministries. If they could completely sever relationships with their original ministries, they could possibly act more independently of them.

While evaluating the nature, organisation and functions of the Planning Commission, it is important to point out the role of historical accident and the personal element, especially the relationship of the top leadership of the Planning Commission with the Prime Minister, whose likes and dislikes as well as trust and confidence played an important role in shaping the course of events. It is remarkable that delineation of the functions and the organisation of the Planning Commission were left to the leadership of the Commission itself (Deputy Chairman and members). It was in the first or second meeting of the Cabinet immediately after independence that the functions of the Commission were approved by the Cabinet. The Cabinet had barely any time to give any thought to the administrative structure of the country or to the detailed functions of the Planning Commission.The composition and functions of the Commission were approved without any serious discussion or debate. If the Planning Commission had been constituted not in January 1972, at the moment of euphoria immediately after independence when the entire economic and administrative system was barely in functioning order, but in 1973 or 1974, it was very likely that a Planning Commission with a different set of functions and a different composition of its top leadership would have been appointed.

The top professional leadership consisted of economists only. It was considered anomalous by many that, in an interdisciplinary organisation such as the Planning Commission, there were no scientists or engineers or other social scientists in the leadership. This was largely due to the fact that the Prime Minister selected the leadership of the Planning Commission from among those who were his closest advisers during the liberation struggle. Most of his closest specialist advisers were drawn from academic life; those among his advisers who were not economists were assigned to other important jobs where their specialist skills were vital. In the Prime Minister's perception the Planning Commission was more a job for economists than any other social or natural scientists. If he did not have persons with training in other disciplines in whom he had trust and confidence adequate enough for appointment in the Planning Commission, he was too willing to leave his economic advisers the task of hiring what technical expertise they needed to help them. In course of time there was considerable pressure on the Prime Minister to change the composition of the leadership of the Planning Commission by appointing persons from the administrative service as well as from other disciplines.

The fact that the functions of the Planning Commission were not exhaustively debated by the Cabinet appears in retrospect to have been a serious error. As the Planning Commission started to function in relation to the bureaucracy, on the one hand, and the political leadership, on the other, it was increasingly clear that there was resistance against the acceptance of the role of the Planning Commission in the way it was originally conceived. This was not to suggest that if the scope and functions of the Planning Commission were fully debated and all the conflicting views were aired, they would have been either fully understood or if understood, would have been readily accepted by the rest of the government and the political leadership. The full implications of many of these functions would have been clear only as the actual planning process unfolded. Since the politicians had little or no experience of running a planned economy and since the administrators and bureaucrats were young and inexperienced, the full implications of the discipline and restrictions imposed by the requirements of planning on freedom of action would have been appreciated only through experience. Nonetheless, the ministers and the administrators or bureaucrats had the distinct feeling that the role and composition of the Planning Commission were imposed on the system by the overwhelming, and unchallenged, authority of the Prime Minister.

(c) The Planning Commission and the Political Leadership

The Deputy Chairman, who was the executive head of the Planning Commission, was not a politician. He was not a member of the party nor was he a member of the Parliament. This was equally true of the members of the Commission. The political element in the deliberations and recommendations of the Planning Commission was to be supplied by the Minister in charge of Planning, who was in the first year or so the Minister of Finance and subsequently the Prime Minister himself. However, the working procedures of the Planning Commission were such that the Deputy Chairman and Members of the Commission did not have regular meetings with the Minister in Charge of Planning so that all political implications of the recommendations of the Planning Commission could be systematically thrashed out. This would also have accelerated the political education of the Commission. The Deputy Chairman and Members did not attend the meetings of the party where most of the divergences and differences of views amongst the politicians were aired. In view of the fact that one party dominated the Parliament so completely, the debates on policy issues took place in the internal meetings of the political party itself.

In fact, the Planning Commission never had a full-time political leader or a politician member of the Cabinet involved in its deliberations. The Finance Minister, when he was the Planning

Minister, had his hands full with the affairs of the Finance Ministry and was unable to devote any appreciable time to the Planning Commission. The Prime Minister became the Minister of Planning and Chairman of the Planning Commission to accord a higher authority and status to the Planning Commission. In view of the inter-ministerial co-ordinating role of the Planning Commission, which imposed discipline on the freedom of action of the ministries, the Planning Commission appeared as a "super ministry" and, therefore, it was considered that this image of the Planning Commission could be made acceptable only by the leadership of the "super minister", i.e. the Prime Minister. In view of the very limited direct involvement of a political leader in the deliberations of the Planning Commission, it remained a purely "technocratic" body. But no set of recommendations could be made which did not have differential impact on different interest groups. No recommendations could in this sense be politically neutral and avoid value judgement. The Planning Commission did not always offer alternative sets of recommendations on any specific issue, with detailed analysis of the pros and cons of alternative proposals in terms of the declared objectives of the political leadership. Most of the time it offered one set of recommendations, after taking into account the divergent views of the ministries. The Prime Minister often implicitly assumed that the Deputy Chairman and the Members were not only fully aware of the hopes, aspirations and objectives of the party and fully shared them, but also that they understood the limitations of action inherent in a political process where differing and sometimes opposing interest groups had to be accommodated in order to preserve the unity of the party and its continuance in power. Not being involved in the political process and in the party's deliberations, the Planning Commission did not fully appreciate the political feasibility constraints on economic policy-making and therefore could scarcely be in a position to weigh the political costs and benefits before offering its recommendations.[2]

The Planning Commission did not actively seek support among the individual members of the Cabinet or in the party leadership for its policy recommendations. It could conceivably seek out the groups in the Cabinet or in the party which would muster support for its recommendations at the time of deliberations in the Cabinet or persuade the Prime Minister. The Finance Minister, when he was the Planning Minister, did lend his political support to the recommendations of the Planning Commission but did not actively canvass support. However, the Finance Minister had powerful enemies in the Cabinet. His effectiveness was further eroded when in the wider political sphere he fell out with the Prime Minister. Even when he was the Minister in Charge of Planning the Prime

Minister, not having been involved in the formulation of the proposals of the Planning Commission, felt no particular commitment towards them. He treated the policy proposals of the Planning Commission as those of the Deputy Chairman and Members of the Commission and did not identify himself with the Commission or with its recommendations. This might have been different if he had participated in the deliberations and in the formulation of the Planning Commission and had given his support to the proposals before they were submitted to the Cabinet. This could have created problems of a different sort, however. Because of his overwhelming authority, a proposal which received his approval was as good as accepted by the rest of the Cabinet.

The Planning Commission's failure or reluctance to win allies within the Cabinet for the support of its policy proposals was partly due to its assumption that in actual decision-making nobody excepting the Prime Minister really mattered; the ministers did not reflect in fact any powerful interest groups and most, if not all, of them individually depended for their political survival on the support of the Prime Minister. This attitude on the part of the Planners appeared, in retrospect, to be a miscalculation because even though it was the Prime Minister who dominated the decision-making process, the decisions had to be implemented by the individual ministers. The latter could still delay or nullify the effects of decisions made at the top. Moreover, granting that the ministers were not very effective spokesmen of the various interest groups and that they did not have the ability to significantly affect the decisions of the Prime Minister, the fact remained that there were interest groups in the party and in the country which exercised pressures on the decision-making process. Even though the Prime Minister did not appear to rely exclusively on any group or any person for advice in decision-making, he, however, discussed important policy issues whenever he felt it necesary with many groups of people ranging from the individual ministers to the individual civil servants, as well as various individuals inside and outside the government and the party, who in his judgement did reflect the views of the various interest groups. This process of consultation and deliberation was conducted in an informal, ad hoc, and often on a bilateral basis. The discussions in the Cabinet or in the National Economic Council were only a part of the process of deliberation. Even though an important forum for public debate was the Parliament, where the various interest groups were represented, it seldom debated at length economic questions of national importance. Moreover, with virtually no opposition members, the possibility of the Parliament engaging in serious debate on the government's policies and proposals was limited. The behind-the-scenes debates or deliberations in the party meetings were more important but they were infrequently held and were not open to the public.

Faced with this process of decision-making, the Planning Commission could have attempted to win support for its policies among the different interest groups in the party or could have resorted to public discussions in the press or the other news media. However, it did neither, partly because it cast itself in a non-political role and partly because of the feeling that the Prime Minister, its closest political patron, would not favour such an activity. Their only means of influencing decision-making was, therefore, to convince the Prime Minister by having separate and private dialogues with him. There were many of those but they were not basically different from the various other bilateral dialogues which the Prime Minister engaged in. The process of decision-making was not institutionalised in a way which would have allowed the different views of policies or programmes to be systematically recorded, co-ordinated and resolved through debate and discussions, especially in the later years.

The method of decision-making, described above, was aggravated by a trend towards the centralisation of power, of both executive functions and policy-making in the hands of the Prime Minister. The importance of the deliberations of the Cabinet in the decision-making process declined over time. The need to intervene in the matters which were within the jurisdiction of the individual ministries was considered by the Prime Minister as having been thrust upon him by the inefficiency and lack of initiative of the ministers. On the other hand, the ministers complained that his interference was unnecessary; the degree of interference varied with the relative strength and personality of the individual ministers. Therefore, the rationale and extent of centralisation in relation to matters lying within the domain of the individual ministers was a complicated two-way process. In many cases it was true that the overwhelming authority of the Prime Minister did reduce the degree of initiative and enterprise; the ministers were afraid of making mistakes and hence of being taken to task by the Prime Minister for their errors. Their narrow or, in a few cases, weak independent political base diminished their standing in the party, in the Parliament and in the Cabinet. Moreover, an increasing public complaint of corruption on the part of the political party further eroded their stature *vis-à-vis* the Prime Minister. The chief casualty, however, was the sense of joint responsibility of the Cabinet as well as the process of "learning by doing".

The tendency towards centralisation in decision-making in the hands of the Prime Minister had the effect of increasing the tension between the Planning Commission and the ministers. The Planning Commission was not only directly under the Prime Minister, but also there was a widespread belief that it exercised appreciable influence on the decisions of the Prime Minister in economic matters. There were certain features in its relationship with the Prime Minister which, even though they were not crucial in the substantive matters of policy-making, assumed disproportionate importance in the eyes of the eyes of outsiders. There was constant competition between the ministers, as well as between the civil servants, to seek frequent contacts and proximity with the Prime Minister. The proximity and frequency of contacts were often considered an index of influence in the decision-making process. The leadership of the Planning Commission had informal and personal contacts and the relationship with the Prime Minister dated back to pre-independence days; this contributed to the impression of having considerable influence. Moreover, the Prime Minister used to demonstrate publicly a great deal of personal respect for the academicians leading the Commission. The fact that he granted them the status of ministers—an honour which was usually accorded only to the members of Parliament in a parliamentary democracy—was considered an instance of exceptional treatment. To the politicians, the professional leaders of the Commission were mere bureaucrats with particular skills who did not deserve the same dignity and status as the elected representatives of the people. The granting of such high status distinguished them from the bureaucrats or permanent civil servants and established their priority over the civil servants, in the matter of participation in decision-making. Again, in administrative matters such as appointments, dismissals or transfers of officers to and from the Commission, the Prime Minister showed a much greater degree of consideration than was true for the other ministries. By his occasional remarks he often contributed to the impression that he valued the opinion of the Planning Commission greatly. This could be due to the fact that he wanted it to be known that his policies did rely on the advice of some of the well-known professionals in the country. Or this could be his own way of raising the status of the Planning Commission in the eyes of the rest of the government. The Planning Commission did not attempt to dispel this impression and to clarify the nature of its influence on decision-making. The truth of the matter was that its recommendations, like his consultations with many other individuals and interest groups, constituted only one of the inputs in his decision-making process.

Over the years, however, the apparent influence of the Planning Commission in the decision-making process declined. The decline in the image of the Planning Commission in the eyes of the public and the politicians accompanied the deterioration in the economic circumstances of the country. The worsening economic situation shown by a deepening food crisis, rapid inflation, drastic deterioration in the balance of payments and an acute shortfall in resources for development, which began in 1973–74, was associated in the minds of many with the failure of economic planning and the incapacity of the Planning Commission to provide correct policy advice and effective solution to the pressing economic problems. In some quarters there was a nagging doubt that the Planning Commission was probably deficient in its diagnosis of the economic problems of the country. The acute foreign exchange crisis in 1974 was traced by some in the government to the inability of the Planning Commission (*a*) to make reliable forecasts or projections of balance of payments and to provide sufficient warning and (*b*) to secure larger and timely economic assistance from abroad. The slow rate of recovery of the economy to the pre-independence level and the sluggish growth in the level of agricultural as well as industrial output and exports were held by many to be due to the lack of appropriate policy recommendations on the part of the Planning Commission.

The criticism of the performance of the Planning Commission in the press and in public discussions strengthened the forces within the government which opposed the acceptance of the Planning Commission's preeminent role in economic policy-making. The most widely publicised criticism of the Planning Commission was voiced by the professional economists of the country in the First Annual Conference of the Bangladesh Economic Association in 1974. The professionals outside the government found the major policy recommendations of the Five-Year Plan out of line with the socio-political realities of Bangladesh in several ways. In the first instance, the government was not thought to be serious about the socialist transformation of the economy; the policies of nationalisation enabled a few privileged groups to enjoy the benefits of state ownership and control. Secondly, the Planning Commission paid inadequate attention to the influence of corruption on the economy and on policy-making in the country; it was guilty of overestimation of the ability of the economic and administrative system to recover from the ravages and dislocations caused by the war of liberation. Thirdly, there was public criticism about the losses and inefficiencies in the public sector enterprises and about shortfalls in domestic resource mobilisation. Both professional and public opinion tended to associate directly the failure in economic performance with the failure in economic planning and with the deficiencies in the capability of the Planning Commission.

The Planning Commission had argued that there were drastic, unfavourable changes in the external economic environment confronting Bangladesh shortly after its emergence as an independent state. The world-wide inflation, which affected most severely the essential commodities like food, fuel and fertilizer, etc., a substantial fall in the terms of trade, fluctuations in agricultural output caused by variable weather culminating in devastating floods in 1974, had all contributed to the economic malaise. Admittedly, there were severe shortfalls in the implementation of reconstruction and development programmes which could be traced to deficiencies in the administrative ministries and executive agencies. These deficiencies were of a nature which did not permit quick solution, though at the time the Planning Commission underestimated the deficiencies and overestimated the possibilities of improvement. In its annual economic reviews and in Annual Plans, it did analyse these shortcomings and did suggest remedies; many of them were not accepted by the Cabinet and the rest of the government or, if they were accepted, could not be or were not implemented.

However, it should be noted that frequently delays occurred in arriving at decisions partly because the decision-making process was unsystematic and partly because there was centralisation in many areas of administration. Moreover, even when decisions were taken in time, there were lapses in implementation. The absence of widely accepted institutional procedures for inter-ministerial co-ordination of policies and activities had contributed to the generally low level of efficiency in administration. Within the socio-political constraints of its foreign policy and external relationships with the major donors, Bangladesh did receive substantial foreign assistance. But there were bottlenecks in the disbursement of foreign assistance for which the Planning Commission as well as the administrative ministries shared responsibility.

(d) The Planning Commission and the Bureaucracy

The relationship between the Planning Commission and the bureaucracy was a function partly of the Commission's relationship with the political leadership and partly of the relationship between the political leadership and the bureaucracy. Before independence, the structure of the civil service or bureaucracy was dominated by the elite core of the bureaucracy, which was called the Civil Service of Pakistan. This particular service had the top positions of authority reserved for them in the administrative structure of the country. This was a "generalist" service and specialised or professional training was not required in principle for the various functions they were to perform. They worked in all the ministries or departments at the top of the administrative hierarchy, irrespective of the specialised nature of the ministries concerned, and could be moved indiscriminately from health to telecommunications or from education to power development sectors. The members of the professional and technical services, i.e. doctors, engineers, accountants, etc., occupied positions inferior to them in the administrative hierarchy. Moreover, there were other categories of "generalised" services which performed similar functions but at the level of the provincial and local governments. The top positions in the latter as well were reserved for the members of the former Pakistan Civil Service. The recruitment to this elite service was carried out on the basis of a single point of entry at a young age and promotion was based on seniority except at the highest level of responsibility where selection was made on the basis of merit. There was a considerable amount of tension and conflict between this particular "elite" service and the other services because of their domination in the administrative structure of the country, especially the near monopoly enjoyed by them in all the jobs of highest responsibility. In the post-liberation period, there were new tensions added to the old ones; there was tension between those who participated in the war of liberation and worked with the government in exile and those who had worked with the Pakistani regime. The vast majority of those who did not flee from the army occupation had no choice but to work with the military regime and only a few of them had actively collaborated with the regime. The distinction between those who submitted themselves to authority because of lack of choice and those who were active "collaborators" with the military regime was often blurred.

In the pre-independence period the bureaucracy, especially the elite service, i.e. the Civil Service of Pakistan, had enjoyed a position of pre-eminence in the decision-making process and in executive functions. From 1958 to 1971 the Pakistan military forces in close alliance with the civilian bureaucracy, had governed the country and distributed patronage and privileges in the economic field as well. With the independence of Bangladesh, the political forces came into the forefront; the political party with its diverse interest groups and supporters asserted their dominance and planned to relegate the bureaucracy to a secondary role in the decision-making process. Moreover, thought was given by the political leadership to the reorganisation of the structure of services and administration, which would have reduced the position of pre-eminence of the elite Civil Service of Pakistan over the rest of the administrative services. With this end in view, the newly appointed Administrative and Services Reorganisation Committee recommended unification of all the administrative services in Bangladesh and the elimination of reservation of posts of higher responsibility for any particular category of service.[3]

The appearance of the Planning Commission on the scene with its higher status, extensive functions and a potentially large organisation appeared to the bureaucracy as an indication of the inroad which was being made into the old order and their hegemony. Firstly, it heralded the intrusion of "outsiders". The leadership of the Planning Commission in Pakistan was vested in the Civil Service of Pakistan. The Bengali members of the Pakistan Civil Service had expectations that on the disappearance of the senior Pakistani bureaucrats they would step into their shoes including the top positions in the Planning Commission. A Planning Commission dominated by academic economists at the top and run by professionals and specialists at the bottom, was seen as a threat to the "old order". Secondly, to make matters worse, the Planning Commission had close links with the political leadership. The political leadership was initially anxious to establish its authority over the decision-making process; to some extent it attempted to reduce the supremacy of the elite service by appointing officials from other than the elite services to positions of the highest responsibility. The Planning Commission by its own nature, composition and preference was seen as an ally of the political leadership in the dilution of its pre-eminent role. The Planning Commission was not only an active party to the recommendations of the Services Reorganisation Committee for the unification of the various categories of services but also it advocated a greater participation of technical specialists in the management of public sector enterprises and agencies. It played an active role in the appointment of technical specialists as the heads of the public sector industrial corporations in the aftermath of nationalisation of the industrial and financial enterprises in 1972. These were positions of great power and patronage, involving command over considerable financial resources and employment, which were predominantly held in the past by the members of the elite Civil Service. The Planning Commission argued the case for the appointment of specialists or experts from outside the established Civil Service on two counts. Firstly, they were expected to infuse new attitudes and bring in new and different background and experience. They were likely to be more innovative and prone to experimentation because they had no strong ties to established procedures and practices. Secondly, the members of the Civil Service were young and inexperienced—most of the officers placed at high levels of policy-making had no experience of working in the central government of Pakistan. The appointment of senior and experienced persons from private industry, professions and academic institutions was expected to overcome the deficiency in experience and knowledge.

The feeling of resentment against the Planning Commission was aggravated not only by the conferment of the political status of Ministers on the Deputy Chairman and Members of the Planning Commission, but also by according the status given to the highest civil servants to the Chiefs of the Technical Divisions of the Planning Commission. Bangladesh inherited a very status-conscious society, where the effectiveness of a person in a particular job and his relationship with the other agencies and officials, his equals as well as his subordinates, depended to a great extent upon the formal status he enjoyed in the hierarchy of the government. The Planning Commission fought for and secured very high status for a few of its top professional and technical staff in order to ensure their effectiveness in dealings with other ministries and agencies, including representatives of foreign governments. In spite of this, in the inter-ministerial meetings which the Planning Commission had frequently to convene on the projects and programmes of the various ministers, other ministries were not always represented at an adequately high level, Thus, the opinions expressed by the ministries did not carry the authority and weight which they deserved. It was felt by the Planning Commission that this was an attempt to detract from the effectiveness of granting a high status to the professional and technical staff of the Planning Commission. Throughout the first years of the Planning Commission the status accorded to its personnel was a constant source of friction in dealings between the bureaucracy and the Planning Commission. It was probably a reflection of the resentment of the members of the elite service that the high status of the leadership of the Planning Commission was withdrawn after the change of government in August 1975.

The Planning Commission assembled together in its staff a large number of specialists in diverse fields and the top echelon of the Planning Commission was drawn predominantly from academic occupations. To the rest of the Government the Commission carried the image of an academic institution and its analysis and policy proposals were considered to be academic. They were thought to be insufficiently aware of the problems of implementation in the field, i.e. the attitudes, habits and capabilities of those in charge of implementation at the different levels of the government; while this was certainly true in a few cases, this was neither a serious nor a general problem. But it was true that their initial lack of intimate knowledge of the established bureaucratic rules and procedure, or more importantly, their propensity to violate them, whenever they found the rules obstructing speedy action, did create frictions and bottlenecks in their working relationships with the administrative ministries. This deficiency was remedied over time and was not considered to be a serious problem at any stage, particularly because within the Planning Commission itself there was a considerable number of officials who had worked in and had close familiarity with the rules and procedures of the government departments and ministries. It was noted earlier that the bureaucracy itself had very limited experience and, therefore, shared with the Planning Commission whatever deficiencies were due to the lack of wide-ranging and long administrative experience. In many areas, the problems and challenges were new and past experience was not a very useful guide and, in fact, could become counter-productive. Both the bureaucracy and the Planning Commission faced the common task of "learning by doing". The Cabinet did not function efficiently, as has been mentioned earlier, and its policy-making role as a collective body was gradually eroded over time. However, to the extent that policies or programmes were subject to systematic deliberations in the Cabinet, the administrative ministries and agencies did bring to bear their points of view, both administrative and otherwise. The permanent administrative heads and the chiefs of the executive agencies did actively participate in the discussions. Whether, in fact, all the aspects relating to a particular policy proposal were brought to bear on the deliberations of the Cabinet depended upon how aggressive and efficient the relevant ministries were in advocating their various points of view.

There were other mechanisms through which the inter-ministerial consultations were undertaken by the Planning Commission. With the accumulation of experience over time, the Planning Commission seldom made policy proposals without consultation and discussions with the ministries and agencies involved in the particular policy or programme. However, in the early years such inter-ministerial consultations were not systematically held and all the agencies or departments concerned were always not consulted. This was partly due to the absence of any established "rules of business" in the government, which could clearly specify the areas or subjects in which the individual ministries had primary responsibility, This created tension between the bureaucracy and the Planning Commission. Instances of what the ministries considered as haphazard methods of inter-ministerial consultation undertaken by the Planning Commission were not lacking. In a few cases the Planning Commission consulted the particular ministers in the course of formulating proposals relating to their particular sectors without at the same time conducting a dialogue with the bureaucracy in the ministry. Its consultations with the ministers were guided by the need to apprise the ministers of policy proposals as well as to receive their reactions before the formal discussion by the Cabinet. But the fact that the bureaucracy was left out at this stage of consultation was naturally resented. Again in some cases, the Planning Commission had detailed consultations with the officials of ministries without the full knowledge of the ministers. The officials in many cases did not keep their ministers informed about the discussions and consultations with the Planning Commission but nonetheless made commitments regarding the views of the ministries, which were disowned by the ministers in the Cabinet meetings. There was an ambivalence regarding the extent of delegation of authority by the ministers to their officials. The Planning Commission sometimes consulted the views of the autonomous agencies or public sector enterprises, without participation by the ministries, which were in charge of regulating the agencies and were superior to them in the administrative hierarchy. For example, the Railway Board in charge of implementing the railway development programme was consulted without involving the ministry of communications, which was the regulatory ministry for the railways. This created complications when there were diverging views between the autonomous public enterprises and their regulatory ministries. In some cases the Planning Commission did it in order to overcome delay and expedite matters because the regulatory ministry was found to act merely as a post office for communicating the views of the agency without its direct involvement. The absence of clear demarcation of responsibility between the ministries and the agencies under them, due to the absence of rules of business, added to the difficulties. But the direct consultations between the Planning Commission and the executive agencies were resented by the regulatory ministries; this was regarded by the ministries as undermining their authority, even though no substantive issues were involved. Eventually the Planning Commission used to bring together both the ministries and the agencies under them for inter-ministerial consultation. But to the extent that there persisted divergencies of views between the ministry and its agency, and these divergencies came to the forefront in course of inter-ministerial consultation, the process of decision-making was complicated. The Planning Commission took the side of either one or the other party, depending upon its judgement and analysis of the problem. In either case, there was a tension in the relationship between the Planning Commission and the ministry or the agency, as the case might be.

Over the years the Planning Commission learnt the fundamental lesson that the bureaucrats could both be pressurised into accepting and implementing policies to which they were totally opposed. In the implementation of the particular policies or programmes the bureaucracy had the ultimate power to delay or frustrate the policy proposals of the Planning Commission or the decisions of the Cabinet. It was not only a matter of wilful opposition or obstruction by the bureaucracy but also a lack of enthusiasm or support for policies or programmes. The motivation of the officials in charge of administration was important, especially when the political leadership was not single-mindedly determined on the implementation of programmes and policies. The concept of a disinterested bureaucracy which was neutral politically or between alternative policies and which carried out to the best of its ability whatever policies were formulated by the Cabinet was not something which was easy to realise in practice. In the early years when the Planning Commission co-ordinated the process of inter-ministerial consultations, it did not take any particular care to record for the benefit of the Cabinet the views of the bureaucracy in the administrative ministries wherever they were different, while making its recommendations. It examined the different views of all the agencies or ministries concerned on a particular proposal and submitted its final recommendations to the Cabinet after taking them into account. The bureaucracies of the relevant ministries did, indeed, reopen the issues and repeated their points of view in the Cabinet meetings whenever they felt that their views were neglected by the Planning Commission. Nonetheless they resented the Planning Commission's method of not faithfully recording or adequately directing attention to their views for the benefit of the higher decision-making authorities; this feeling did not improve the general environment of co-operation.

Moreover, the leadership of the Planning Commission did knowingly or unknowingly vaunt on many an occasion an air of intellectual superiority over the bureaucracy. These were undoubtedly occasions when, in their treatment of the bureaucracy, they were not sufficiently considerate or polite. The tension was aggravated whenever they wanted to demonstrate their closeness to and affinity with the political leadership, especially the top political leadership. The Planning Commission did not always fully appreciate that while it was required to evaluate and co-ordinate ministerial programmes and policies, very great care was needed in the way in which it communicated its judgement and opinion about the performance or attitudes of other ministries; the manner in which it conducted itself in relation to other ministries affected its effectiveness in the machinery of the government. Traditionally one of the methods of winning over the bureaucracy was for the technocrats to carry the burden of professional and technical work while letting the bureaucracy take the credit or receive the acclaim for the work. This was the way in which the bureaucracy, at least the upper echelon of the bureaucracy in Pakistan or indeed elsewhere in the sub-continent, has utilised in the past, whenever necessary, the technical expertise of professionals, especially that of the academic professions. The Planning Commission was unwilling to follow this tradition. The fact that the bureaucracy considered the academic economists in the Planning Commission as having usurped the authority and power which had originally belonged to it was resented by the Planning Commission. There was a vicious circle. The Planning Commission had a compulsion to prove that it was not inferior in practical policy-making functions and that the bureaucracy was not really as good as it claimed to be in terms of its expertise in administrative and policy matters. Resentment and arrogance on both sides fed each other in their mutual relationship and led to a decline in the spirit of co-operative action. The fact that by its influence on or persuasion of the Prime Minister the Planning Commission occasionally got certain changes introduced or certain decisions taken, in spite of the opposition of the bureaucracy, did not improve matters. The politicians or the ministers, on their part, did not attempt to bring closer together the Planning Commission and the bureaucracy. In its attempt to exert its authority over the members of the bureaucracy and in the context of its own sense of insecurity, emanating partly from its inexperience in administration, the political leadership was not averse to making use of the underlying tension between the two. As the years went by, the Planning Commission was beginning to be recognised and accepted by the bureaucracy as a necessary adjunct of the government; the degree of co-operation between the two in general improved. The Planning Commission by the nature of its functions and responsibilites continued to create tensions of a different kind. This was the constant attempt to suggest changes in policies and institutions, in procedures and methods, in relationships between the ministries, in respect of their individual functions, in the system of relationships between the agencies and ministries, and so on. To the bureaucracy the Planning Commission often appeared to be pressing for too many changes at the same time; this they not infrequently blamed on the lack of realism and practical wisdom of the Planning Commission. In late 1973, one of the most senior civil servants—a generalist-administrator, who had experience over many years of working in development ministries associated with agriculture—was appointed as a member of the Planning Commission, It was hoped that this would bring to bear directly on the deliberations of the Planning Commission the administrative and practical considerations involved in the formulation of programmes and policies, and also promote a greater co-operation with the Planning Commission and the bureaucracy. In point of fact, this move introduced additional tension between the bureaucracy and the academic planners within the Planning Commission itself and did not help matters.

It is essential for the successful operation of a planning system that the bureaucracy and the administrative machinery of a country as a whole must accept it as an integral part of the decision-making apparatus. Its functions and responsibility must be accepted by the rest of the government. There is an educative process involved; the educational process must encompass all the three parties: the political leadership, the bureaucracy and the planning machinery.

The political leadership and the bureaucracy need to learn and accept the implications of the planning process in the economic development of a country of which the Planning Commission is intended to be the embodiment. The political authorities must provide strong and consistent support to the planning machinery in the performance of its functions and in the realisation of its stated objectives. The planning machinery must share the hopes and aspirations of the political leadership; it must have the same set of convictions. If the political authorities make radical departures from their original objectives and goals, with which the planning machinery is associated, the composition and the leadership of the planning machinery should change along with the change in objectives. As stated before, the political leadership was not totally committed to its announced objectives and goals nor was it determined enough to resolve and overcome whatever conflicts existed between its various interest groups. The planning machinery, which had to formulate policies and programmes for the implementation of the stated goals, could not escape the consequences of vacillation in policies and its reputation in terms of both intellectual integrity and competence could not but suffer.

The planning machinery, on the other hand, must understand the socio-political "feasibility constraints" on economic policies and programmes; it must be aware of what is feasible within the constraints placed by the prevailing administrative machinery, as well as by habits and attitudes of the bureaucracy. There is a dilemma because planning must involve changes in the existing structures and institutions. How to bring such changes about in the context of a peaceful social and political environment is not an easy question to answer. Any change, be it in the field of administrations, institutions, ideology or attitudes, creates tensions and strains. It involves changes in the relative powers of different groups in society as well as in the government. The Planning Commission has the task of guiding and stimulating the rest of the government in accepting the principle of rationality in policy-making, in the context of the socio-political objectives of the political leadership. If it initiates changes and actively sponsors new programmes and policies affecting the rest of the government, while the bureaucracy and the administrative machinery do not appreciate the need for such changes, the planning machinery may face constant resistance and frequent failures. It must persuade and convince; it can put pressure provided it is backed by the political authority, at appropriate times and in appropriate areas. This involves delicate judgement. The Planning Commission must ideally be in a position to play a leadership role. For it to successfully play this role, it must be accepted as such by the rest of the government; if conflict develops between the Planning Commission and the rest of the system the political leadership would need to resolve it, if necessary, by making changes in the administrative system which would enable the planning machinery to play its leadership role. It is not always within the capability or commitment of the political leadership to change the administrative system rapidly and fundamentally to suit the requirements of planning. It could be accomplished very slowly in many cases.

Thus the leadership role of the planning machinery in the system of economic policy-making and development administration can succeed on two conditions. One is that the political leadership must be more closely associated with the deliberations and functions of the planning machinery than was the case in Bangladesh. The top leadership of a planning machinery must not only be technically competent but must also be an integral part of the political leadership. It must be willing to play a political role, with all that it implies. The non-political technical experts can serve in advisory roles in the Planning Commission but cannot provide its leadership without detriment to the effectiveness of the Planning Commission unless they are willing to play a political role. At the least along with the technical experts, there must be active participation by the political leaders in the working of the planning machinery. The second requirement is that so long as the elitist civil service with its exclusive attitudes, rights and privileges is predominant in the bureaucratic machinery and so long as the bureaucracy and the professional and academic occupations in the country exist in separate and distinct compartments, without reciprocal interactions and mobility between the two, it is not easy and indeed may be counter-productive to induct into the leadership of the planning machinery experts or persons who are not part of the top echelon of bureaucracy. Under these circumstances for the sake of cohesion and reducing the loss in efficiency arising out of frictions and hostility between the non-compatible groups, it may be advisable to associate with the leadership of the planning machinery the top echelons of bureaucracy rather than outside experts or specialists, however eminent they may be in their chosen fields of specialisation. If the rest of the administrative machinery thinks, for whatever reasons, that the planning machinery is an academic institution, no matter how hard the academics try to develop administrative acumen, the credibility of the planning machinery in the eyes of those who are in charge of implementation of policies would remain less than what it should be. What the rest of the administrative system or structure thinks of the planning machinery is no less important to its success than what it actually is. If the political leadership feels strongly committed to the abandonment of the concept of an elitist civil service and is determined to draw freely upon the widest spectrum of expertise and experience in the society at all levels of the government and the decision-making process, it must act decisively and introduce deliberate changes in the existing system. It must make its intentions clear by firm and determined action. It must be convinced, in order to do so, that the benefits of such a change outweigh the costs.

An additional factor which tends to reduce the effectiveness of the planning machinery is the existence of ideological differences between it and the bureaucracy. The leadership of the planning machinery, being the political appointees of the government, is expected to be committed to the ideology of the government of the day. If the permanent civil service and bureaucracy is opposed to it, a new dimension of ideological differences is added to aggravate the tension in its relationship with the planning machinery. The concept of an ideologically neutral bureaucracy, conceived in the historical tradition of the Western-style parliamentary democracy, serving in successive governments with conflicting political ideologies, appears to be increasingly obsolete in the political systems of the developing countries, which are in a state of continuous flux. The above are a few lessons demonstrated by the experience of the Planning Commission in Bangladesh in its relationship with the political leadership, on the one hand, and the bureaucracy, on the other.

Notes

1. "Organisation of the Bangladesh Planning Commission"—a Circular of the Cabinet Division, Ministry of Cabinet Affairs, Government of the People's Republic of Bangladesh, dated 31 January 1972, Dacca, Bangladesh.

2. The first chapter of the Five-Year Plan deliberated on the socio-political perspectives of planned development. But this was not the result of a dialogue between the political leadership and the Commission; nor did it represent a consensus reached between the divergent political interest groups in respect of the socio-political requirements of planning. It was more in the nature of an exhortation by the Planning Commission from a group of experts standing outside the political processes or pressures to which the ruling party was continually subject.

3. The Bangladesh Constitution contained no special protection for the Civil Service in terms of their dismissal or removal from service. They could be dismissed without showing any cause and without giving them the right of appeal and defence. Government of Bangladesh, *Report of the Administrative and Services Reorganisation Committee*, April 1973, Dacca, Bangladesh.

3. The Planning Commission and the Administrative Ministries

(a) Introduction

The relationship of the Planning Commission with the political leadership, on the one hand, and with the bureaucracy, on the other, affected its role in the entire governmental machinery for taking decisions and implementing policies. The degree of success it achieved in performing its functions in relation to the other ministries was related importantly to this triangular relationship between the Planning Commission, the bureaucracy and the political leadership. The Planning Commission was assigned to perform three important sets of functions relating to the other ministries and executive agencies in the government. Firstly, it was to allocate resources to the various sectors of the economy, under the administrative control of the different ministries. Resources were allocated not only between sectors but also in terms of individual projects in each sector. While the Finance Ministry was in charge of collecting and disbursing domestic resources, as determined in the Annual Plan and the annual budget, the Planning Commission was in charge of mobilising external resources, i.e. foreign economic and technical assistance. It undertook detailed negotiations for projects and programmes on behalf of the various ministries with the foreign aid agencies, both bilateral and multilateral. Secondly, the Commission was in charge of evaluating and appraising development projects in respect of their technical and economic feasibility. It was only when the individual projects were evaluated and declared to have met the test of social profitability by the Planning Commission, that they could be sent to the Cabinet for approval. Thirdly, the Planning Commission was to formulate and recommend changes in institutions and policies required for the successful implementation of development projects as well as for a general improvement in the efficiency of the performance of the economy.

While the ministries publicly proclaimed the need for planned economic development, there was an ambivalence in their approach towards planning because of the discipline and constraints which a Plan imposed on their freedom of action. The constraints which the planning process imposed on the ministries appeared to them as an indication of an undesirable propensity on the part of the Planning Commission to extend its power or control over the rest of the government (a propensity which was not warranted by the requirements of an efficient allocation of resources). They conceived of the task of planning in terms of allocation of financial resources by the Planning Commission and the Cabinet to individual ministries or agencies, at the beginning of a five-year period or at the beginning of each financial year. According to this view, it was the responsibility of the individual ministries to make intra-ministerial or intra-sectoral allocations and to choose the most efficient combination of projects and programmes within each sub-sector and sector. It was not readily seen that there was an interdependence between projects and programmes in different sectors and that it was the responsibility of the central planning agency to ensure an overall consistency between them.

There were two factors which contributed to this misconception or limited conception about planning. One was the attitude of the politicians and ministers who felt that as representatives of the people, they were responsible to the people and Parliament for the development of the sectors of the economy under their jurisdiction. Therefore, they had the authority individually and, indeed, it was their responsibility to determine the priorities in their respective sectors and to select individual projects and programmes. The fact that the Planning Commission made the recommendations about the inter-sectoral and intra-sectoral priorities, on the basis of objectives laid down in the Plan, as approved by the Cabinet, and that projects and programmes at the stage of implementation, were, in fact, to be approved by the Cabinet, with the ministers participating in the process of approval, did not appear very satisfactory to them. Even though the Cabinet made the final decisions regarding inter- and intra-sectoral allocations, it was the detailed recommendations made by the Planning Commission which provided the basis of their decisions. These recommendations in themselves, even if they were in some cases modified by the Cabinet, appeared as constraints on the freedom of action of the ministers. They would have liked the Planning Commission not to make recommendations relating to projects and programmes within their respective jurisdictions; they would have preferred not to subject their expenditure programmes firstly to evaluation by the Planning Commission and secondly to collective scrutiny by the entire Cabinet.

The second factor which led to a misunderstanding of the role of the Planning Commission was the attitude of the officials in the various ministries. The bulk of the officials of the government of Bangladesh belonged to the pre-liberation provincial government of Bangladesh (former East Pakistan) and had no experience of or familiarity with the functions of the Planning Commission of a national government. The Pakistan Planning Commission located in Pakistan mainly worked in relation to the federal or central ministries of the government of Pakistan. In fact, the image of the Pakistan Planning Commission, in so far as the former government of East Pakistan was concerned, was that of an agency which was used by the Pakistan regime to discriminate against East Pakistan, to justify smaller allocations of resources and to obstruct the development of East Pakistan. The Bangladesh Planning Commission insisted that in the changed economic circumstances all projects and programmes of the various ministries, even though they were evaluated before independence, tended to bring back memories of the obstructionist tactics of the Pakistan Planning Commission. This was especially so since some of these were approved after a great deal of effort, persuasion and political pressure as worth while, even by the government of Pakistan.

The role and the functions of the planning process and the constraints and discipline imposed by the requirements of planned development would probably have been more readily understood or appreciated by the individual ministries, if they had their own planning cells to do systematic thinking on their own priorities and objectives. The logic of overall planning in terms of the relationship between objectives and instruments, including the trade-off between competing objectives, would then have been brought home to them. The existence of planning cells in the individual ministries would have simplified the task of educating and initiating the ministries in the concepts and techniques of planning. This was not to say that there would not have been disagreements and conflicts between the Planning Commision and the planning cells in the ministries on the determination of priorities and on the selection of projects. But then there was a much greater likelihood that the process of communications would have been more informed and effective. The absence of planning cells in the individual ministries was partly due to a lack of appreciation of the need for such cells and partly to the shortage of trained manpower. The individual ministries were greatly in need of trained staff to undertake (*a*) project preparation, (*b*) review and supervision of development projects on a continuous basis and (*c*) evaluation of the performance of their respective sectors. In fact, the ministries ended up with a bundle of programmes or projects without in any way systematically thinking through their priorities. Each ministry pressed for larger allocations without considering the effect on the other ministries or without suggesting measures for increasing the total resources for investment. In view of the scarcity of resources, their programmes had to be drastically curtailed. The Planning Commission made recommendations about the curtailment of the programmes of different ministries to the

Cabinet. In Cabinet discussions the ministries were unwilling to suggest the curtailment of each other's programmes; in other words, they fought shy of determining inter-sectoral priorities by collective decisions. Each was afraid to suggest reductions in the programmes of the other ministries or more particularly, to suggest the inter-sectoral allocation of the reduction in the overall programme, lest others were displeased. Each one wanted the Planning Commission to propose the curtailment in the programmes of others; thus, they sought to lay the responsibility squarely at the door of the Planning Commission.

The Planning Commission was involved in at least three different kinds of inter-ministerial co-ordinating functions. One was the co-ordination of the inter-ministerial consultations for the formulation of policy proposals or recommendations to be sent to the Cabinet. These policy proposals related to the individual sectors of the economy, but they had implications for and affected the policies in the other sectors of the economy. Second was the co-ordinating function relating to formulating Annual and Five-Year Plans in terms of a consistent set of programmes and projects within the limits of the available resources and in conformity with the development objectives laid down in the Five Year Plan. Third was the co-ordination of the activities of the various ministries or agencies in connection with the implementation of the development projects or programmes. This function was mainly intended to remove major bottlenecks in the way of implementation. Therefore, it was selective in nature and it was undertaken either at the suggestion of one or two ministries involved which needed co-ordination or by the Planning Commission when it identified major bottlenecks in the course of its evaluation of the progress of implementation of the Annual Plan or the development projects.

(b) The Planning Commission and Policy Co-ordination

In each of these three types of co-ordinating functions, the Planning Commission was entering terrain which had possibilities of both conflicts and co-operation with the ministries. Firstly, on the issue of the co-ordination of inter-ministerial deliberations for the formulation of policies, questions were raised as to whether the Planning Commission would undertake the inter-ministerial co-ordination itself or would just provide its advice or opinions on particular policy proposals. In the latter case the actual co-ordination was to be done by the particular ministry which originated the policy proposed or by the sectoral ministry to which the particular policy proposal related. It was also suggested as an alternative that the actual co-ordination of the inter-ministerial deliberations could be undertaken by the Cabinet Secretary, or by the Prime Minister's office which, not being responsible for any particular sector or policy, was expected to be neutral and could ensure that all the differing points of view were reflected not only in the deliberations but also in presenting the recommendations to the Cabinet. The Planning Commission's participation in the deliberations was accepted as necessary in the sense that it would examine the various points of view and could express its opinion, taking into consideration the effects of a policy proposal on the overall economy and on inter-sectoral relationships and consistencies.

It was argued on behalf of the Planning Commission that all policy proposals on economic matters such as trade, exchange policy, industrial and agricultural policies, etc., should be co-ordinated by the Planning Commision itself. This was because the Planning Commission, it was pointed out, was formally conceived as a secretariat of the National Economic Council. The latter was to meet more frequently than the Cabinet in order to concentrate on more detailed deliberations on economic issues than was possible for the Cabinet as a whole, all the more as the cabinet in the course of time grew very large—consisting of about twenty to twenty-five ministers. The Prime Minister, as the head of the National Economic Council and as the Minister in Charge of Planning, would be personally involved in the deliberations of the Council. All the ministries were to send their policy proposals in the first instance to the Planning Commission, which would then consult with and obtain the views of all the relevant sectors and ministries. On the basis of inter-ministerial consultations and its own analysis, it would then put up the final set of recommendations to the Cabinet for their action. In other words, under this scheme of things, no policy proposal would go from any other ministry to the NEC/Cabinet except through the Planning Commission and after the Planning Commission had formulated its recommendations on the particular policy proposal. The ministries considered this to be a surrender of their right to put up their own policy proposals to the Cabinet; they were willing to consult the Planning Commission on their own volition provided they were convinced that the Planning Commission had something to contribute to any policy proposal.

For the Commission to insist that all proposals on economic policy matters must go through it and that it was solely responsible for producing the final recommendations to the cabinet, appeared to the ministries an unnecessary demonstration of its "superior" role which they resented. While from the point of view of efficiency in economic policy-making, it was unnecessary for the Planning Commission to press for this particular procedure for conducting inter-ministerial deliberations or for bringing the views of the

Planning Commission to bear on the formulation of the economic policy, it was at the same time shortsighted on the part of the individual ministries to insist that they would decide at their discretion whether the Planning Commission should or should not be consulted, before it sent its proposals to the Cabinet, and that it was for them to decide whether they would bring the overall perspective provided by the Planning Commission to bear on the formulation of economic policies. What was important was that the Planning Commission should be given the responsibility of examining a policy proposal in relation to the overall economic perspective and in terms of its economic merit and its effects on the other sectors of the economy, and that the results of its examination should be fully considered in inter-ministerial deliberations.

The viewpoint and perspective of the Planning Commission in analysing specific problems in the general context of the economy and the overall objectives of the Plan should have been an essential input in the deliberations of the Cabinet, the latter might also be expected to take an overall view, distinct from that of particular ministries, as did the Planning Commission. To illustrate, we can take a simple case such as the fixation of the price of sugar cane. The pricing policy for sugar cane affected tax and subsidy policy, the optimum pattern of sorting, the relative role of modern and cottage industries, and a whole host of related issues. Every year at the beginning of the growing season, the government was to announce its purchase price; it determined the former's choice between sugar cane and competing crops, such as tobacco, for example. The large-scale manufacturing industry which was nationalised was the largest purchaser of sugar cane. The purchase price offered by the industry also affected the grower's decision as to how they would divide their sales between the modern, larger-scale industry, on the one hand, and the indigenous cottage industry on the other. It also affected the incentive to smuggle sugar cane across the border to India. The choices between alternatives were to be evaluated in the light of the efficiency of resource use and the possibilities of international exchange open to Bangladesh. The distribution of output between the indigenous and modern industry had income distributional effects as well as impact on the revenues of the government. The factory-produced refined sugar was wholly distributed under a system of public rationing. If the price of sugar, distributed under the rationing system, was below the cost of production, determined largely by the price of sugar cane, there was the need for a subsidy which was related to the broader problem of raising resources for financing it. It was a moot question whether under the conditions of extreme poverty prevailing in Bangladesh, sugar was an essential item of consumption deserving a subsidised price. Furthermore, to the extent that refined sugar was consumed in the more prosperous urban areas and the indigenous sugar in the rural areas, a lower price for refined sugar compared to sugar produced by cottage industry was inequitable. To the extent that domestic output was supplemented by cheaper imports, part of the proceeds from sale of imports could be used to finance the subsidy on the sale of domestic output.

Finally, as a result of long, drawn-out discussions between the Planning Commission and the Cabinet, it was decided that the list of policy issues on which the views of the Planning Commission were to be sought would be specified. In fact, they were closely linked with the subjects listed as matters for deliberation and decision by the National Economic Council. In all these matters the ministries, before putting up proposals to the Cabinet, were required to consult the Planning Commission. There were occasional lapses in the consultation procedure, sometimes deliberate and sometimes by mistake, which led to tension between the Planning Commission and the ministries. The Planning Commission necessarily attended all the meetings of the National Economic Council. The meetings on inter-ministerial consultations were held by the ministry directly concerned with the subject under consideration, and the recommendations were drafted by the ministry concerned; in case the views of the Planning Commission were different, they were reflected separately in its report submitted to the Cabinet. However, in a few cases, when the views of the different ministries were sharply conflicting, the ministries themselves voluntarily asked the Planning Commission to hold inter-ministerial consultations. The Planning Commission's mediation was considered useful by those ministries which were relatively weak in terms of staff strength or of ministerial leadership. They were often willing to let the Planning Commission do the staff work, hold inter-ministerial meetings, and draft the final set of recommendations, provided they were sent in the name and with the authority of the ministry rather than the Planning Commission. Occasionally in the case of policies or proposals towards which they understood that the Cabinet, especially the Prime Minister, was not favourably disposed, they were willing to have the Planning Commission take the initiative in making recommendations and thus, face the consequences of his disapproval.

The tension between the Planning Commission and the ministries had another aspect. This was related to the initiative taken by the Planning Commission on various policy matters on which the ministries were not very enthusiastic but on which the Planning Commission felt strongly and pressed for changes and improvement. This can be illustrated with reference to two policy matters which created considerable controversy. One related to the organisation and management of nationalised industries, as discussed in a later chapter, and the other related to the institutional arrangements for the preparation of investment projects which cut across the jurisdiction of several ministries. In these matters the Planning Commission often produced recommendations after detailed inter-ministerial consultations but they ran into difficulties because of either opposition or indifference of the ministries. It was learnt through experience that policy prescriptions or recommendations for changes which did not arise from the felt needs of the ministries or sectors were not likely to be accepted and, if forced upon the unwilling ministries, could be defeated in their implementation. The ministries had to be persuaded that the changes were in their interest and until this persuasion was complete and conviction dawned, no amount of cabinet resolutions was to be of much avail,

(c) The Planning Commission and Project Preparation and Evaluation

The attempt to establish a project preparation unit or cell in the Planning Commission was considered by the ministries as an outright usurpation of their legitimate functions. The successful implementation of the Five-Year Plan and its investment programmes depended on the availability of a substantial portfolio of projects. As the rehabilitation and reconstruction phase immediately following independence was over, the foreign aid agencies were pressing for an increase in the proportion of project assistance to commodity assistance. Negotiation for project assistance took a long time, including the time involved in the appraisal and evaluation of projects. If a large pipeline of projects could be made ready, the process of negotiations for project assistance could start early, so that every year an adequate number of projects would be ready for implementation. This would allow an even or regular flow of commitment and disbursement of project assistance over a number of years. The foreign aid agencies had repeatedly pointed out that in the absence of well-prepared projects, the process of aid negotiation and hence the flow of assistance was likely to be slowed down. The Planning Commission, being in charge of negotiating foreign assistance, was under great pressure to produce aid-worthy projects. It did put pressure upon the ministries to improve their capability for project preparation; it negotiated and secured foreign technical assistance to improve and supplement domestic capability for project preparation.

Two limiting factors in the process of project preparation were recognised by the Planning Commission. First was the dearth of an adequate number of technically trained personnel in the ministries. Second was the fact that there were a number of projects which cut across the jurisdiction of several ministries; that there was no institutional mechanism for co-ordinating the various ministries and agencies in the preparation of these multi-sectorally linked projects. Prominent amongst these were such projects as (*a*) integrated rural development projects and (*b*) water control and irrigation projects. The rural development projects were composite projects consisting of (*a*) rural credit, (*b*) agricultural extension service, (*c*) co-operative development, supply of agricultural inputs such as seeds, fertilizers and pesticides. etc., (*d*) rural public works programme, etc. Three ministries were principally involved: those concerned with rural development, co-operatives and local government. The projects had to be prepared jointly by the three ministries; the various components were interdependent, and consistency had to be ensured between the various components. Similarly, water control and irrigation projects involved not only the ministries of water and agriculture but also fisheries and inland water transport as well as roads and railways. The irrigation projects were divided between two ministries, i.e. the water ministry was in charge of (*a*) large- and small-scale gravity flow (canal), surface irrigation projects and (*b*) deep tubewell irrigation projects. It was also in charge of large- and small-scale flood control projects. The agriculture ministry was in charge of low-lift pump, surface water irrigation projects as well as ground water irrigation such as deep and shallow tubewell projects. These projects were in many cases interrelated; they competed for the same sources of water and in a few cases they were supplementary to each other. The depth and width of canals for "gravity flow" irrigation determined the availability of water for low-lift pump irrigation from surface water. The preparation of each of these projects, especially projects which had elements of both irrigation and flood control, required close co-operation between the different ministries. Since they affected agricultural production, it was necessary to relate the projects to the requirements of water for different cropping patterns and intensities. Similarly, water control projects affected the availability of water, both the amount of flow and its timing or seasonality, for the development of fisheries. They also affected the system of drainage, which in turn had an impact on the flow of water for navigation. Therefore, water control projects had to be prepared while keeping in view their repurcussions on such sectors as agriculture, fisheries and inland water transport, as well as the requirements of these sectors themselves. In view of the heavy dependence of road and railways on river bridges, the amount and level of water flow was also important for railways and road communications.

The Planning Commission suggested that these projects which had closely related and interdependent components, which belonged to a number of ministries, and between which it was essential to ensure consistency should be formulated under the aegis of the Planning Commission. It suggested the same procedure for those projects which, although they did not have components belonging to different ministries, affected a number of sectors in various ways. The Planning Commission was to bring together experts and specialists from the various ministries or agencies, together with its own experts in the related fields and appropriate foreign technical experts and consultants, for the formulation of such interlinked projects. It was to co-ordinate the diverse points of view of the relevant sectors, to provide a forum and leadership for the resolution of divergent and conflicting viewpoints.

An alternative approach would have been for one Ministry to take the leadership in co-ordinating and conducting different ministries; moreover, it could appoint subject-matter specialists belonging to the various sectors to enable them to take into account all the diverse aspects of the projects. For example, the Water Ministry could have agricultural, fishery and navigation experts to examine the related aspects of water control projects. There were two limitations to the latter approach. First, there was a dearth of competent experts in various fields so that each ministry could not build up project preparation units consisting of experts in so many diverse disciplines. Secondly, consultation and co-ordination with related ministries were not easy since differences of opinion between the experts of various ministries were often substantial. It was necessary to have neutral experts who belonged to none of the executive ministries in order to mediate between them. There was always an element of uncertainty in the appraisal and evaluation of conflicting or divergent viewpoints and consequently, there was room for judgement. The possibility was relatively small that one of the executive ministries could effectively co-ordinate the other related ministries as, for example, would be necessary in the case of rural development projects with components relating to different ministries.

The Planning Commission was relatively neutral as between different ministries in the evaluation of the different aspects of projects. In the case of projects which did not have multi-sectoral components but had serious inter-sectoral repercussions, the ministries other than those which were in charge of the projects concerned were more willing to accept the co-ordinating role of the Planning Commission. The Water Ministry rejected the suggestion that the Planning Commission might act as a co-ordinator in the case of water control and irrigation projects, in order to ensure an adequate consideration of inter-sectoral repercussions at the stage of project preparation, as amounting to a surrender of its primary function and responsibility.

Although the Planning Commission was to assume a co-ordinating role, the principal function in the preparation of projects in terms of technical and economic feasibility studies was to be performed by the ministry and its field agencies; the inter-ministerial project preparation unit or cell under the leadership of the Planning Commission could at best bring together the various ministries for an intensive cross-checking of assumptions, facts and diverse technical opinions bearing on the projects. It could act as a catalytic agent, stimulate and expedite the preparation of projects by ironing out differences and ensuring that all the aspects of a project were considered. Past experience indicated that inadequate attention was paid to all the relevant aspects of projects when left solely to the responsibility of the ministry in whose principal jurisdiction such projects fell. The involvement of the Planning Commission and its technical personnel at all stages of project preparation could confer one great advantage in that it would shorten the time and expedite the process of appraisal and evaluation of a project by the Planning Commission for its inclusion in the Annual or Five-Year Plan.

A project preparation cell was indeed established in the Planning Commission with this end in view; a technical assistance loan from the IDA/IBRD was obtained for the purpose of securing technical expertise and consultants, both domestic and foreign, to help in the tasks of project preparation mainly, but not exclusively, in the field of water development and irrigation projects. But its success in accomplishing what it set out to do was very limited. The persistent opposition and non-co-operation of the ministry prevented it from co-ordinating the preparation of water projects. Since nearly all the technical input in terms of data and preliminary studies and investigation of such projects had to come from the ministry, such an enterprise could not succeed without the co-operation of the ministry. Eventually it was agreed that the ministry would undertake the preparation of such projects, with whatever help from the Planning Commission they needed in the process, and would itself consult with and co-ordinate with other related ministries.

In its task of preparing and co-ordinating the Five-Year and Annual Plans, the Planning Commission attached critical importance to the evaluation and appraisal—cost–benefit analysis—of individual development projects. This was a function assigned to it by the decision of the Cabinet. But as the rules and procedures for co-ordination and appraisal were formulated by the Planning Commission, there was considerable resistance on the part of the ministries to using them. The question whether the Planning Commission should have the responsibility at all of appraising and evaluating projects, before they were considered suitable for inclusion in the plan was debated. The ministers, as politically responsible heads of their ministries, thought that for them to surrender the task of project approval to an expert body like the Planning Commission was to surrender their principal function in the field of economic management and development. The Planning Commission argued that while the projects were to be formulated in the first instance by the agencies in charge of implementing the projects and the respective ministries should undertake the initial cost-benefit analysis, the ultimate responsibility for evaluation and appraisal should lie with it, as the central co-ordinating agency of development projects. The ministries, on the other hand, held that the projects should be sent directly to the Cabinet for final approval. While they were agreeable in principle to seeking assistance from the Planning Commission, if necessary, in the task of project appraisal or analysis, they argued that it was the ministries and not the Planning Commission which had the necessary level and range of technical skills in their respective fields to appraise projects. The Planning Commission felt that its role was, nonetheless, important in two respects. Firstly, in respect of financial and economic analysis of the projects, it was not possible for each ministry to build up its own expertise. There was in any case a serious shortage of economic expertise in the executive ministries for the work of project appraisal. Hence, the use of a centralised pool of economists in the Planning Commission, specialised in project appraisal work, appeared to be the most efficient way of performing this important task, In addition, the Planning Commission was to formulate common criteria for project appraisal such as the estimation of "accounting prices" for various inputs which were to be used by all the ministries in the technical evaluation of projects. Secondly, the Planning Commission argued that it should have a second look at the technical aspects of the projects as well; this could be done with a much smaller component of technical skills than that of the executive ministries and agencies. A ministry which sponsored a particular project and was in charge of its implementation had vested interests in it, and was not eager to examine alternative techniques of production; it tended to produce one blueprint. An external agency like the Planning Commission was, in this sense, "neutral". It had a critical role in requiring the ministries to re-examine the projects, persuading them to think of alternative blueprints and cost estimates based on a different set of assumptions regarding factor mix, etc.

Moreover, it was argued that the co-ordinating role of the Planning Commission and the centralisation of the function of project appraisal in its hands did not imply that ministries and agencies responsible for the formulation and implementation of projects were not closely associated with the task. In fact, the Planning Commission would place the results of its evaluation and appraisal before an inter-ministerial committee (Inter-ministerial

Project Evaluation Committee) composed of the representatives of the administrative ministry, including those of the agency in charge of implementation, and the representative of the Ministry of Finance. The Planning Commission was to serve as the secretariat of this inter-ministerial Committee. The experts of the Planning Commission and those of the ministry or agency concerned would have prior discussion on the various aspects of a project before its examination by the Committee. Similarly, the Ministry of Finance was to examine the financial aspects of a project, involving in particular, such aspects as: (*a*) recurring expenditure, (*b*) salary structure, (*c*) personnel requirements and staffing pattern, (*d*) costs of materials, (*e*) any other matter having financial implications.

The findings of the inter-ministerial committee were to be merely recommendatory; its report was to be submitted to the National Economic Council for approval. It was only when the National Economic Council approved a project that it was declared economically and technically sound. Furthermore, unless a project was included in the Annual Plan/Budget, mere "approval" in respect of its economic feasibility, was not a sufficient condition for any ministry to incur expenditures on it. Whether a particular project was included in the Annual or Five-Year Plan depended on available resources and the alternative use of resources between the competing sectors and projects.

As a result of extensive discussions in the Cabinet and the National Economic Council, the Planning Commission's role in the evaluation and appraisal of projects, as outlined above, was agreed to by the Cabinet and the ministries. However, a compromise was struck in that the ministries should have the authority to grant the final seal of approval about the feasibility of projects below a certain size. All projects costing Tk.2.5 million or less were to be left to the administrative ministries for appraisal according to the standards laid down by the Planning Commission. The delegation of authority in respect of small projects would also reduce the burden of work of the Planning Commission. However, as a safeguard against mistakes or incorrect use of the criteria prescribed for project evaluation, the results of evaluation by the ministry were to be sent to the Planning Commission for their comments, though not for approval. The Planning Commission prepared a project evaluation manual which indicated in detail the data required, methods of examination of data, and techniques of cost–benefit analysis, which were to be followed by the ministries. The Planning Commission did not have the right to reject such projects. If the comments of the Planning Commission conflicted with those of the ministry at the levels of officials, they were to be referred to the minister in charge whose decision would be final. The Planning Commission felt that this was a compromise reached in order to allow the ministers authority to approve a few projects on their own. even if not judged economically and technically worth while, on extra-economic grounds. It was considered necessary in order to secure the co-operation and agreement of the Cabinet and the ministers to the entire procedure for project appraisal as outlined above. If the ministries followed the procedure, as laid down by the Commission, in most cases this should not have been unsatisfactory. There was, however, one danger in this decentralisation. There was the possibility that a ministry, in order to avoid the scrutiny of the Planning Commission, would split up a project into sub-projects, each costing less than Tk.2.5 million, and this would tend to frustrate the objectives and spirit of the evaluation procedures. In fact, there were a few occasions in the following years when this apprehension was justified. Moreover, there were cases when the adverse comments by the Planning Commission on projects below Tk.2.5 million, which were independently approved by the ministries, were ignored by the latter; there were also cases where schemes costing less than Tk.2.5 million were not sent in fact for comments to the Planning Commission, but were approved by the minister.

The misunderstanding of the role of the Planning Commission in the task of project evaluation was partly due to the inadequate realisation of the various steps in project evaluation and their interrelationships. These were: (*a*) projection evaluation by the Planning Commission and by the inter-ministerial committee; (*b*) approval by the Cabinet of the viability of a project; and (*c*) inclusion of a project in the Five-Year and Annual Plans. Even after a project was approved by the Cabinet as technically and economically feasible, it was not realised that this did not ensure the inclusion of the project in the Annual or Five-Year Development Programme, since this depended on the resources available. The ministers assumed that once a project was considered viable, they could go ahead and start expenditure on the projects. The resistance on the part of the administrative ministries to the project evaluation procedures set up by the Planning Commission was partly due to what the ministries considered unnecessary and "difficult to meet" criteria, in terms of information and analysis required by the project evaluation manual. Moreover, their manpower and expertise were not sufficient to meet the "high standards" set up by the Commission. This was particularly true in the social services and agricultural sectors, where work on project preparation was continually lagging behind. It was felt that the criteria set up by the Planning Commission could not easily be used for projects in the social sector where information on output and benefits was not easily available or quantifiable. The Planning Commission eventually modified the data requirements and rigours of analysis in order to adjust to the prevailing shortcomings in the capability of the ministries and in the data base. It was better to have a simplified analysis rather than to insist on a level of analysis which required expertise and data not available in practice and which was wrongly done or not done at all.

In terms of actual expenditures on development projects, the Planning Commission exercised great leverage. Since most projects had their foreign exchange component financed by foreign assistance, the Planning Commission which was in charge of negotiating and allocating foreign assistance could determine the expenditures on projects. In the case of commodity assistance, the Planning Commission was not only involved in its allocation to the public sector agencies and the ministries for meeting their requirements of current inputs, but also determined, in co-operation with the Ministry of Foreign Trade, foreign exchange expenditures in the private sector.

Control over the negotiation and allocation of foreign economic assistance added to the image of the Planning Commission being a "super ministry"; in particular it created tension with the finance ministry. These functions, combined with the control over the allocation of domestic resources to development projects as part of the Annual Development Programme, left the Finance Ministry basically with three sets of functions: (*a*) tax policy and administration and management of domestic public debt, (*b*) control over non-development expenditures, both domestic and foreign currency, and (*c*) control and regulation of the banking system. This was considered by the Finance Ministry a significant diminution of its traditional powers, especially when the broad category-wise allocations of foreign exchange expenditures, including foreign exchange expenditures in the non-development budget, were brought under the general supervision of the Planning Commission. The Planning Commission was involved in advising on the allocation of foreign exchange to broad priority areas; allocations to sectors and ministries had to be agreed by it. Since a major part of the current "non-development imports", i.e. raw materials, etc., was financed by foreign credit, the Planning Commission was involved jointly with finance and foreign trade ministries in determining priorities for its use in the course of negotiating with foreign creditors. In practice, it took the leadership, which was resented by the Finance Ministry. The Finance Ministry felt that it should be the sole agency responsible for supervising and controlling the use of the inflow and outflow of foreign exchange resources, as part of the process of preparing a budget for overall resources, foreign as well as domestic. The realisation was slowly to dawn that so long as the Planning Commission was involved in determining the priorities of the investment programme as well as in influencing the composition of current output in the substantial public sector, its association with the allocation of foreign exchange was a logical extension of its functions.

(d) Review of Economic Performance and Supervision of Projects

The Planning Commission, as a part of its task of formulating Annual and Five-Year Plans, evaluated the past performance of the various sectors of the economy. Periodic evaluation of the performance of the economy involved the progress of implementation of the development programme in each sector. It not only reviewed the growth of output in various sectors, but also the utilisation of inputs in scarce supply, including foreign exchange, which were allocated under a system of licensing. The activities of all the ministries came under review in the process of evaluation. It included not only the ministries which were in charge of implementation of development projects but also ministries involved in more general macroeconomic functions such as the Finance Ministry and Foreign Trade Ministry, involved in implementing tax and foreign trade policies. The ministries had the feeling of being scrutinised by the Commission which generated a feeling of hostility on their part. In its analysis and evaluation of shortfalls in the achievement of targets, the Commission was to identify and analyse the reasons for such short-fall, and to suggest remedies for improvement in performance.

The Planning Commission had within its administrative control the central statistical organisation, which was an essential adjunct to the planning process for the collection and dissemination of statistics relating to all the sectors of the economy. The Commission felt that in view of the very unsatisfactory data, the statistical organisation should be strengthened and the collection of data in the various ministries should be co-ordinated and centralised wherever feasible in the interest of efficiency and economy. It was illustrative of the general atmosphere of suspicion, mistrust and the image of the "super ministry" which the Planning Commission evoked that the suggestion for the reorganisation and strengthening of the statistical organisation under its aegis received stiff resistance from the ministries, including the finance ministry. Its recommendations were to build up an expanded central statistical organisation, to be headed by a high-level expert enjoying the status of the highest ranking permanent civil servant and to be placed under the Planning Commission for the purposes of administrative control and guidance. More specifically, this implied (*a*) raising the status of the statistical organisation in the administrative set-up and increasing the degree of autonomy in terms of budgets and personnel, (*b*) expanding its functions and personnel and (*c*) integrating three other statistical agencies located in different ministries with the Bureau of Statistics. The most important amongst the independent statistical agencies which were to be integrated were (*a*) agricultural statistical agencies and (*b*) population census organisation. Integration was advocated for the sake of economy, to avoid a duplication of functions, especially in view of the dearth of highly qualified manpower in this field. The statistical cells in the various ministries engaged in collecting data related to the daily operations of the ministries were to be maintained; they were to receive technical help and guidance from the central organisation, which would strive to secure a greater uniformity and comparability of and an improvement in the quality of data collected by the various ministerial statistical cells. A National Statistical Council was to be constituted, in which all the user agencies as well as ministries, collecting specialised, operational data of their own, would be represented. The administrative ministries objected to the integration of statistical services in one central agency, primarily because it meant a diminution of their power, but they were eventually persuaded to accept it, largely because the finance and the establishment ministries (the latter concerned with personnel and recruitment matters in all the ministries) accepted the argument of economy and optimum use of manpower. But they opposed the case for greater autonomy in budgetary and personnel matters for it implied a loss of their own power.

There was also strong opposition to the suggestion that such an integrated and high-level organisation should be placed under the aegis of the Planning Commission. The alternative suggestion was that it should be placed in the office, and not under the Planning Commission on the ground that the latter, being the largest potential user of the statistical services, would influence the objectivity and quality of data to suit its own needs, and perhaps would also attempt to cover up the deficiencies of the Plan. It was not clear why the Planning Commission should have more of a vested interest in publicising the satisfactory performance of the economy than the Cabinet and the political leadership of the country; it was not an institution independent of the rest of the government. The success or otherwise of the plans and policies of the government basically reflected the activities and performance of the various ministries and the entire government. If anybody had a 'stake' in 'doctoring' or 'producing' the data, it would have been the political leadership as a whole or the individual ministries engaged in the implementation of specific policies and programmes. The possibility of political interference with the collection and presentation of data could not be ruled out but this could not have been solved by changing the location of the statistical organisation within the government; basically it was a matter of autonomy and independence of the statistical organisation. Moreover, there would have to be independent checks by means of critical examination of data by universities, independent research institutes and scholars. These methods had worked fairly satisfactorily in the pre-independence period. A statistical organisation in a close working relationship with the Planning Commission under the guidance of a National Statistical Council, with the relevant ministries, non-governmental individuals and institutions represented on it, would have considerable autonomy in budgetary, administrative and personnel matters. It seemed to be a reasonable solution to the problem of improving the standard of statistical services. It would have provided the necessary safeguard against political interference while ensuring at the same time technical competence as well as close co-operation with the functions of economic planning and policy-making. Three years of negotiation and sustained argument were required before the principle of integration was accepted.

In its evaluation of the performance or progress of specific development projects, the Planning Commission had to identify bottlenecks and shortcomings in a way which often reflected unfavourably on the administrative ministries. As an illustration one may mention its evaluation of the irrigation programme, which pointed out the slow progress of the irrigation projects and inefficient or inadequate utilisation of irrigation capacity already in operation.[1] This was primarily due to the inefficient institutional and managerial system as well as inappropriate pricing policies. A lack of co-operation amongst the officers belonging to different ministries and working at the local level on the various aspects of the irrigation programme, an unsatisfactory training and extension programme, as well as inadequate maintenance and repair facilities which were pointed out in its report as a few important causes of inefficiency, constituted a castigation of the performance of the ministeries in charge of the irrigation programme. The above example was a clear indication of how and why the Planning Commission's function of evaluation of progress and performance of the economy in general and development projects in particular placed it in a position to "monitor" performance within the machinery of the government. Inevitably this strained its relationship with the different ministries.

It was the Planning Commission's involvement with foreign aid negotiation which brought it into direct contact and involvement with the day-to-day administrative and implementation problems. Increasingly the Planning Commission was involved in monitoring the disbursement of foreign assistance; its success in negotiating additional foreign economic assistance each year was dependent on the rate of disbursement of past assistance. Even though it was not fully equipped to monitor, supervise and evaluate the progress of implementation of foreign-aided projects, it assumed an active role in reporting and identifying bottlenecks. In many cases it also made recommendations about specific steps to be taken for the removal of bottlenecks and for speeding up implementation. Starting in 1974, the Planning Commission began to organise meetings of the relevant ministries concerned with the particular aid-financed projects. These meetings which sometimes also included the local representatives of the aid missions of the foreign governments were intended to identify the bottlenecks, determine the particular agencies which could remove bottlenecks and the actions which were required from the various ministries for the speedy implementation of development projects. Since almost all the development projects had an aid-component, very few projects escaped this process of scrutiny and review. The inter-ministerial meetings for the review of the progress of aid-financed projects involved the Planning Commission in keeping pressure on the ministries to expedite such administrative aspects of project implementation as the appointment of consultants and of engineers, placement of international tenders, procurement of materials, customs regulations, payment of taxes on foreign-aided goods, and transportation of equipment and machinery to the site of the project, etc. The periodic review of the progress of aid-financed projects was welcomed by the donors; in so far as the ministries were concerned, the effect on the relationship of the Planning Commission with the ministries was mixed. The ministries which suffered from non-co-operation or delays on the part of other ministries welcomed whatever help such meetings could render in the removal of bottlenecks. The ministries which were identified as responsible for delays or lapses in execution did not particularly appreciate such review meetings. However, since all the ministries were engaged in the implementation of projects, and since each one had at various times both faulted on its part and suffered because of inefficiencies of others, it was expected that on the whole and in the long run, all the ministries would welcome the coordinating and monitoring role of the Planning Commission, even though in the short run such a role could not but generate frictions.

Notes

1. This is analysed in detail in the last chapter. "Decreasing Command Areas and Performance in Minor Irrigation Programmes"—unpublished report, 10 October 1974, Planning Commission, Dacca.

4. The Planning Process: The Five-Year Plan

(a) Introduction

The formulation of the Five-Year Plan was one of the few areas of economic policy-making to which the government publicly committed itself. The Election Manifesto in early 1973 laid emphasis on the formulation of the Five-Year Plan in order to "lay the foundations of economic development in Bangladesh". In his first address to the newly elected Parliament in 1973 the President of Bangladesh referred to the Plan[1] as a means of realising the cherished dream of economic emancipation which inspired the nation to make so much sacrifice for attaining political freedom.

There were some both inside and outside the Planning Commission who thought that the preparation of a Five-Year Plan should be delayed and that the Planning Commission should concentrate on the preparation of the rehabilitation and reconstruction programmes for the war-damaged economy. This was not done because it was argued that rehabilitation and reconstruction programmes could best be formulated in the context of a long-term development programme. Moreover, the task of formulating concrete short-term policies and programmes for the reconstruction of the economy would be performed by the Annual Plans, the instruments through which the Five-Year Plan was to be implemented. Since the transition from reconstruction to development was a gradual one, the exact point in time at which reconstruction faded into development was a matter of judgement. The Prime Minister frequently mentioned three years as the period of reconstruction during which his government could not be expected to provide any improvement in the living conditions of the masses—a period which would be marked by privations and sacrifices owing to the ravages caused by the war of liberation.

The scarcity of resources in relation the requirements of the reconstruction and development programmes (including the ongoing projects inherited from the past) made it imperative to devise a rational and consistent framework for investment decisions. Disciplined, systematic and hard thinking about alternative policies and institutions was needed urgently. Immediately after independence significant changes had taken place in the pattern of ownership in trade, industry and finance. A well thought out institutional framework was required to ensure efficiency and to promote initiative and enterprise. The preparation of a Five-Year Plan was expected to facilitate the process of such analysis and decision-making. Another important, though unpublicised reason for the preparation of the Five-Year Plan was its use as a basis for negotiating foreign economic assistance. There was a widely-held impression that the donors expected a recipient country to have a Plan setting out a consistent framework of investment priorities and policies, which would enable them to assess the needs for assistance and the form it should take.

The evolution of the views of the political leadership of the party on the broad aims and objectives of planned development in Bangladesh has been discussed in earlier chapters. No specific guidelines about targets and priorities were provided by the political authorities beyond those general guidelines adumbrated in various public pronouncements, election manifestoes and finally. in the constitution of the country. It was left to the planners to formulate concrete and quantitative objectives and targets and to work out their implications in terms of the size and composition of investment programmes, the resources required to finance them and the appropriate policies and institutions needed for implementation.

(b) Planning Models

Since planning has gained in importance in developing countries, it has become usual to use analytical and statistical cum mathematical models as an aid to plan formulation. Bangladesh, however, used a relatively simPle model or analytical framework for the preparation of a first Five-Year Plan and it is of some interest to examine the contribution that it made to the formulation of the Plan and even more to the administrative process of reconciling policy objectives and achieving consistency in the Plan itself. There are inevitability severe limitations to the use of such models in the circumstances of Bangladesh and their use goes hand in hand with all the other activities that are inseparable from plan preparation. Nevertheless the model used proved itself to be a powerful tool in investigating the implications of alternative versions of the Five-Year Plan and in bringing home to ministries the need to coordinate their activities.

The Plan (1973–78) was not formulated in the light of a long-term perspective plan. There was no time to formulate such a plan and it was scarcely possible to utilise previous tentative attempts to formulate a "rough brush" perspective of the long-term development possibilities of Bangladesh.[2] The most sophisticated of these studies explored the long-term prospects for the Bangladesh economy in connection with the development of the agriculture and water resources of Bangladesh. The model sought to show the investment needed to achieve high levels of employment up to the year 2003 and the foreign assistance that would be entailed. It was a highly aggregated model covering only seven economic sectors.

The core of the model consisted of a series of equations relating per capita GNP to the shares of the individual sectors in the GNP. These could be used in conjunction with an independent agricultural study to work out the implications of alternative rates of growth of agricultural output on the rest of the economy. The agricultural study used an optimising model to show the rate at which agricultural output should be expanded, given constraints on the availability of water and land, the demand for various agricultural commodities and implementation capacity for the agricultural development projects. The growth rate of GNP was found to be highly sensitive to changes in the growth rate of the value added in agriculture; a change in the latter from 3.9% to 4.5% per year was associated with a change in the growth rate of GNP from 5% to 6%—an elasticity of greater than one. Interesting though the model was, there were doubts about some of its basic assumptions such as changes in productivity in the various sectors of the economy, capital output and labour output ratios and considerable additional work and testing of assumptions were necessary before any use could be made of it in the preparation of the Plan. It had, in fact, little application to the preparation of the Five-Year Plan.

The model used by the Planning Commission for the preparation of the Five-Year Plan was of an entirely different character. It was in fact a simple input–output consistency model. The Plan was prepared in a year, and there was neither time nor manpower, nor adequate data to attempt anything more sophisticated. It was suggested at one stage that the preparation of the Plan should be postponed until adequate data on all aspects of the economy were collected. The fact remains, however, that whether or not there was a Plan or an economic model, economic policy decisions had to be taken on the basis of whatever data were available. Five-Year Plan or no Plan, there would be annual budgets involving decisions on mobilising resources, development and reconstruction expenditures and so on—which would involve commitments for the future.

The impetus provided by the need to formulate a Plan acted as a spur to the collection of data. It gave a sense of urgency to the organisation of systematic work on monetary statistics, national income accounts, price statistics, foreign trade and cost of living statistics, including the preparation of an annual balance sheet of supply and demand for essential items of consumption such as rice, edible oil, sugar and textiles. The Planning Commission became for a while a depository of all essential data for policy-making; it gathered together data from the various agencies and checked them for consistency; it urged the agencies to examine their own data more thoroughly and to collect additional information through surveys, wherever possible. The use of the input–output model in the preparation of the Five-Year Plan involved a check on the consistency of data emanating from the different administrative ministries as an integral part of the exercise.

(c) The Five-Year Plan Model

The five-year planning exercise used a thirty-three sector conventional input–output model as a means of checking consistency between final demand, sectoral outputs and sectoral allocation of investment resources; the estimation of investment requirements included working capital as well as depreciation allowances for existing capital stock. The sector classification of the input–output table was not entirely satisfactory. Nineteen out of the thirty-three sectors used were industrial sectors. This was due to the greater availability of more detailed industrial statistics. In the agricultural sector only a few sub-sectors were distinguished but they included the major sectors of rice and jute in addition to the fishery, forestry and livestock sectors.

The exogenous variables used included targets for public and private consumption in 1977/78, i.e. the terminal year of the Plan, for exports in the same year. In addition, investment programmes derived from independent studies of ten sectors of the economy, including such important import substitution activities as the production of rice, petro-chemicals and fertilizers were exogenously determined.[3] Such important import substitution activities as the production of cement, cotton textile and steel products were, however. treated endogenously. The Plan aimed at self-sufficiency in cement and cotton textiles, and substantial reduction in imports of iron and steel products. An analysis of the possibilities of producing commodities efficiently in Bangladesh determined the magnitude and pattern of other imports and the results were incorporated in the input–output model.

The reasons for treating investment decisions exogenously varied. In the case of sugar, paper and petro-chemical production, it was partly due to the fact that investment decisions had already been taken and partly that in the petro-chemical complex, a large discontinuous and lumpy investment with substantial economies of scale could not be handled within the static framework of the input–output analysis. The latter set of considerations was particularly relevant in the case of investment in gas lines and power generation.

In the field of agriculture, the rice production programme was treated exogenously mainly because different levels of output required different technologies and, therefore, involved indifferent input–output coefficients. Investment programmes for flood control, health, population and education were also treated exogenously mainly because they were determined by socio-political considerations and little was known about the input–output coefficients involved. In the education sector, for example, there was no analysis of manpower requirements by skills for different levels of output and final demand. Investment in flood control projects during the Plan period was not strictly related to the agricultural target of the Plan, because many of the projects would not be completed within the Plan period and those that were would yield output only after 1978. In all cases, the implications of the exogenously determined levels of output and investment for the rest of the economy were worked out with the help of the model.

Investment in capacity expansion in the various sectors was considered for the five-year period as a whole, on the assumption of a certain rate of growth of investment expenditures during the Plan period. However, the results were not sensitive to changes in the rate of investment during the Plan period. Appropriate lags were assumed between investment in new capacity and the flow of output from it. Investment expenditures during the Plan period were related to changes in output, between the bench-mark year and terminal year of the Plan period, by means of incremental capital coefficients.

In addition to investment expenditures, the plan included other types of expenditure directed to development. The main items were interest payments during the construction period, some subsidies, distribution costs of agricultural inputs, costs of land, and expenditures on wages and salaries during the period of construction and before the completion of investment projects.

The Plan took specific account of non-monetised investment. A villager often made investment in the repair and construction of his dwellings using his own labour and materials, which he either produced in his homestead or on his farm, or collected from his fellow villagers through barter or common sharing agreements. The farmer or village trader built godowns or stores for his produce or merchandise, repaired embankments and maintained his own farm. Investment of this type was generated in the non-monetised sector and did not pass, except very marginally, through the monetary system or the financial institutions.

The input–output model yielded estimates of investment requirements for the terminal year together with forecasts of domestic savings; it provided also an estimate of the balance-of-payments gap. Estimates of import requirements were made for investment goods (related to total gross investment) in raw materials (related to sector outputs) and consumptions goods (related to total private and public consumption). Broad assumptions were made about the way that trade would evolve over the Plan period but it was expected that variations in imports and the inflow of foreign capital from year to year in response to needs, would help alleviate any serious imbalance between supply and demand. Large fluctuations in prices would not arise as a result.

In projecting the balance-of-payments gap for the entire Plan period, it was necessary to consider at least two additional factors which were not incorporated in the input–output model: the impact of delays and uncertainties in the completion of development projects, and the consequences of fluctuations in weather on agricultural output, especially food. Additional foreign exchange was needed to provide for this, but how much would be required could only be judged rather arbitrarily on the basis of past experience of shortfalls in performance and fluctuations in output. In fact, the provision was made on the side of inadequacy, as subsequent developments indicated.

Five alternative investment programmes of different sizes were worked out for the five-year period, reflecting mainly, but not entirely, alternative rates of growth of GNP and consumption demand that might be attempted.[4]

The target consumption level used in the model was intended to provide a minimum basket of consumption goods, including such items as rice, clothing, edible oil and sugar. The consumption target took account of desirable increases in per capita income, past trends in consumption, and minimum socially acceptable norms of consumption for certain essential items. The projections of private consumption, therefore, involved elements which were matters of policy as well as of forecast. The composition of the consumption basket was very relevant for the choice of production techniques, and it affected the choice between large and small-scale production as, for instance, in the case of cotton textiles and leather goods.

The employment implications of the various sizes of investment programmes were worked out on the basis of sectoral capital–labour ratios, which in turn were related to the choice of techniques and scale of production. Capital–output coefficients were closely linked with the capital–labour coefficients. A more detailed exercise outside the model was undertaken for estimating the employment implications of the agricultural programme, the principal components of which were modern inputs such as improved seeds, fertilizer and irrigation.

It is pertinent to recount a number of the limitations of the model. Firstly it was developed in terms of 1972/73 prices. The following years were characterised by violent changes in both absolute and relative prices as a result of upheavals in world prices and domestic inflation. The inter-sectoral coefficients, which were expressed in constant 1972/73 prices, were to change markedly as a result of drastic changes in relative prices in the following years. The changes in relative prices affected the pattern of efficient import substitution very considerably. In the new circumstances, different production possibilities and import patterns deserved consideration. Secondly, the technical coefficients used in the model assumed that the productive capacity in the economy would be operated as intensively as in 1969/70 and in some cases at a much higher level.

The level and structure of final demand and the need for additional capacity during the Plan period was geared to these assumptions but as subsequent events demonstrated, this was an over-optimistic assumption and the coefficients which actually determined production were far from the full capacity coefficients. Nothing could have demonstrated more conclusively that the input–output coefficients were not a matter of just production and the composition of output; they depended initially on the efficiency of organisation, management and institutions.

The increase in the agricultural output from the extension of irrigation facilities depended critically on the efficient use of water. This in turn depended on whether the farmers' groups made the maximum use of their facilities, on whether the extension services and the training of farmers resulted in the efficient use of water, and on the provision of suitable incentives to improve efficiency, including the establishment of remunerative prices for inputs and the sale of output.

Similarly, better utilisation of industrial capacity depended as much on an improvement in managerial capacity as on an enlarged supply of raw materials and intermediate inputs.

(d) The Use of a Model in Plan Formulation

The first Five-Year Plan prepared for Bangladesh was not, of course, the first that had been prepared for the area. When Bangladesh (former East Pakistan) was a province of Pakistan, five-year plans had been drawn up for the region as part of Pakistan's Five-Year Plan. There were two such plans with separate regional components in the decade between 1960 and 1970. Past attempts at planning in Bangladesh had provided many occasions for analysis and study of the overall strategy and broad priorities of development. The results of these studies, although they had been undertaken in a very different political and economic context, still served as a background for future planning efforts.

The model was not mechanically applied to formulate either the size or the sectoral priorities of the Five-Year Plan, but the results of its analysis and the information it yielded were useful in a number of ways. It was an invaluable aid in main judgements and in checking those of others; it was of assistance in the critical examination of the results of various independent studies on the investment and resource potentials of Bangladesh. In formulating the Plan it helped in working out the broad orders of magnitude of sectoral outputs and investments in response to various levels of final demand. This also enabled an examination to be made of the implications for the rest of the economy of alternative magnitudes of exogenously determined investments. The total amount of investment which was exogenously determined in the selected sectors constituted about 30% of the total investment in fixed and working capital. The model also was useful in working out the foreign exchange requirements of alternative sizes of the Plan. This was particularly important because foreign exchange was a crucial constraint. The model indicated the orders of magnitude of the saving–investment as well as balance-of-payments gaps involved in Plans of various sizes. This provided the basis or the occasion for discussions of the political and economic issues involved in varying degrees of dependence on foreign aid.

Excessive dependence on foreign aid was not politically desirable; moreover, there was no assurance that a large amount of foreign assistance would in fact be available. Reliance on a large and uncertain amount of foreign aid was likely to subject the Plan to a considerable degree of uncertainty. Given the availability of domestic resources, these considerations imposed an upper ceiling to the size of the Plan. There was also a floor below which the size of the Plan could not fall, if it was to raise per capita income, provide minimum amounts of essential consumption goods and reverse or at least arrest the growth of unemployment. The problem of achieving a politically acceptable rate of growth, within the amount of aid considered acceptable or likely to be available constantly plagued the planners and political leadership in Bangladesh. These constraints placed a heavy burden on the mobilisation of domestic resources, in an endeavour to reconcile conflicting objectives.

The size of the Five-Year Plan which was finally chosen, as the exercises of the model indicated, required a greater inflow of foreign aid than was found acceptable, partly on non-economic grounds. Consequently, a lower level of dependence on foreign aid was incorporated in the Plan. This implied in turn an upward adjustment in the rate of export expansion or import substitution and hence a greater rate of domestic saving. This was obviously unrealistic but was an example of the way in which compromises in the presentation of the Plan had to be made for political reasons. There were, however. two reasons advanced in justification of incorporating in the Plan a degree of dependence on foreign aid lower than was in fact realistically anticipated. Firstly, a lower foreign aid target was expected to stimulate efforts at domestic resource mobilisation. Secondly, it was expected that if in fact there was a shortfall in domestic resources and if an increased amount of foreign aid was available to offset the shortfall, a larger foreign aid would be accepted; in other words, dependence on foreign aid *ex post* would be larger than *ex ante* and this was found to be politically more acceptable.

The sectoral allocations derived from the exercises performed by the planning model, provided a framework for the ministries or agencies to work out their sectoral programmes. Since the preparation of the sectoral investment programmes or of the broad sectoral priorities was undertaken simultaneously with the preparation of the overall Plan, there was a constant feedback between the two. The final size of the Plan and its broad sectoral allocations were arrived at in a series of iterative steps, which consisted of successive consultations with the ministries or agencies concerned with the sectoral programmes.

The estimates of available resources and the consequential limit to the size of the Plan derived from or revealed by the model had implications for the discussions and consultations that the Planning Commission held with the individual ministries and executive agencies on the size of the sectoral development programmes. The combined total demand for resources by all the ministries was greatly in excess of the availability of aggregate resources which the most optimistic forecasts of the planning model indicated. The original sectoral programmes submitted by the various ministries to the Planning Commission also were in all cases in excess of the sectoral allocations which emerged out of the macro-economic exercise. These allocations in fact acted as a ceiling for the sector programmes so that the demands from the ministries had to be pruned. This process forced them to examine their projects more closely and eliminate non-essential components. When projects and schemes were not fully worked out, which was frequently the case, the ministries were asked to indicate their costs and benefits on a preliminary basis. This imposed a discipline in the task of examining the size and priorities of the sector programmes.

The great educative value of the model, however, was in persuading those who were in charge of sectoral programmes to think in terms of inter-sectoral relationships. The inter-sectoral consistency and co-ordination required by the model brought the sponsors of sectoral programmes to confront each other, not only as competing but also as complementary claimants of resources. For example, the model was helpful in persuading the ministries in charge of power, transport and communication, of the need to relate investments in these sectors to levels of output in the directly productive sectors. Proposals for the location of a project in any sector in a particular region had to be justified either with reference to the infrastructure already available or with reference to the plans for expansion of the physical infrastructure in the region. It was similarly necessary to justify the programme for investment in transport and communications in terms of the requirements for these services by the rest of the economy. The assumptions regarding the future growth of other sectors were crucial in determining the size of their own programme. Investment for the generation of power, for example, had to be consistent with the growth of output in the industrial sector or with requirements for electrification, let us say, of irrigation in the agricultural sector.

The highly aggregative input–output model did not and could not yield allocations for all the sub-sectors or individual projects. The degree of aggregation was much higher for the non-industrial sectors, but even in the industrial sector disaggregation was far short of what was necessary for working out a detailed list of projects. The results obtained from the model had to be supplemented by the following guidelines which were provided for the ministries and agencies engaged in the programmes or projects within their respective sectors:

(1) When considering priorities within each sector, it was necessary to consider the inter-sectoral interdependence and co-ordination of the various sector programmes.
(2) Every effort had to be made to choose techniques of production and organisation which saved on scarce inputs like capital and foreign exchange and substituted labour, wherever possible.
(3) Each sector should prepare a "hard core" minimum programme on the basis of the following considerations:
(*a*) projects must yield output or results during the Plan period;
(*b*) they should have immediate impact on the generation of employment;
(*c*) they should save capital and foreign exchange to the maximum extent possible;
(*d*) the hard core programme should exclude ongoing or new projects which could be postponed until the next Plan period without detrimental effects on the projects to be undertaken during the current period;
(*e*) maximum possible staggering of projects should be undertaken, wherever possible.

In the actual determination of sectoral projects and programmes the interrelation between what it was proposed to do first and what could be done in the next plan period could not be ignored. Projects for flood control and irrigation were a case in point. Because of the long gestation lag in many such projects, investments made in the first plan period (especially those in the latter part of the period) were likely to yield results only in the next plan period or later. It was necessary in a few cases to select projects of this nature within the current period, lest the possibilities of meeting future needs were completely sacrificed. Exclusive concentration on short-run projects, it was feared, might diminish unduly the possibility of productive investment and income growth in the next plan period. Similarly, it was decided not to postpone all import-substitution projects, which would not yield benefit in the current plan period because, given the strictly limited possibilities of export expansion in the immediate future, such a policy was likely to aggravate the balance-of-payments problem in the next plan period.

There were several independent checks on the sectoral allocations, apart from what was derived from the results of the input–output analysis. One was to consider the previous shares of the various economic sectors in the investment programme in pre-independence Bangladesh. Another consideration was that certain sectoral programmes had reached a critical momentum, which should not be lost. The most recent levels of investment programmes in the various sectors, appropriately corrected for price changes, provided a plausible lower limit to the investment programmes during the Plan period. Moreover, there were a few sectors which had traditionally suffered from a lack of funds. This was true particularly of the social sectors like health, education and social services; these sectors had also suffered in the past from a lack of adequate preparation and feasibility studies. Moreover, they had also been handicapped by having a weaker organisational and institutional framework for the implementation of projects than those in the directly productive sectors such as industry, power and transport and communications. The Five-Year Plan attempted to compensate for past neglect of these sectors by increasing the proportion of development expenditures allocated to them. It was assumed that strenuous efforts would be made to increase their absorptive capacity.

The inter-sectoral competence of the administrative ministries in the formulation of their sector programmes, which differed widely, also affected the inter-sectoral allocation of investment resources. In a few cases the ministries took the initiative in preparing their investment programmes and in persuading the Planning Commission to accept the size and priorities of their sectoral plan. This was particularly true of such ministries as agriculture, rural development, water resources and flood control and power development. They had a higher state of preparedness in terms of projects, mainly owing to studies and analysis of their development potential undertaken in past years. In other ministries such as industry, transportation and communication, housing and physical planning, education and social services, past analysis and studies relating to their investment priorities were limited and initiatives in formulating projects and programmes frequently lay with the Planning Commission which played a more active role.

Mention has already been made of the constraint placed on the freedom of choice of investment priorities and projects by the existence of the ongoing projects. There was a large number of ongoing projects at various stages of completion. There were some projects on which very little actual work had been done except that machinery and equipment had been ordered; there were others for which the process of evaluation and appraisal was completed but work had not been started. At the other extreme, there were projects which were awaiting "trial runs" of production before they were made fully operational.

Because of the scarcity of resources, the problem of choice had two aspects: first, there was the problem of choice from amongst the ongoing projects themselves and secondly, there was the problem of choice between the ongoing and the new projects. In a few cases the ongoing projects were complementary to each other. For each ongoing project a number of alternatives could be considered: (*a*) scrap and realise the value of the assets by selling them off; (*b*) complete it according to original design at the optimum speed; (*c*) complete it but at a slower pace; (*d*) complete it but after redesigning and pruning the scheme; or (*e*) preserve or "mothball" the asset already created—with a view to resuming work in the next Plan period. It was ideally possible to work out for any ongoing project the discounted value of the stream of costs and revenues for each of the alternative courses of action indicated above, the objective in each case being the maximisation of the discounted sum of excess of revenues over costs, given the choice of a rate of discount. However, if a project had to be finished in different speeds and forms, working out the stream of costs over time for the remainder of the period would raise a host of problems, not the least of which would be the availability of data in necessary detail.

The basic rationale of this approach was that bygones (as the sum of costs incurred) were bygones and that one was concerned with the comparison between (*a*) return yielded by the project (when completed either modified or at a different speed than originally planned) over the costs incurred in completing it and (*b*) the revenue from selling off what already exists of the project. However, the application of this criteria in dealing with the ongoing projects was subject to at least two limitations. First, there was an interdependence between individual projects, i.e. between the ongoing projects themselves as well as between the ongoing and the completed projects. This complicated the calculation of revenue or benefits from the completion of the ongoing projects, at varying speed or in varying forms. Second, often the ongoing projects were financed by "tied" aid. They were financed by a particular donor country which supplied the equipment and machinery and undertook the task of construction. The option of scrapping the project would have involved one of three alternative courses of action: loans could be forgiven or cancelled by the creditors; the creditors would take back the assets; the creditors would allow the sale of assets anywhere else so that Bangladesh would receive the sale proceeds but would continue to bear the burden of debt, irrespective of the amount of sales proceeds.

Before taking any action in relation to the ongoing projects it was necessary to obtain the agreement of the donor country, and probably in addition, the agreement of the construction company and suppliers of equipment and machinery. In practice, it was found extremely difficult to change, let alone scrap, the design and composition of an ongoing project undertaken under tied project aid. Delays in the execution of a project did occur either due to the shortage of resources in Bangladesh for complementary or supporting local expenditures or due to the inability of the supplier to meet the delivery schedules. However, the donor countries or suppliers seldom, if ever, withdrew or agreed to delay a project in response to the needs of Bangladesh, or to reorder priorities, or to rephase the completion of a project. In fact, speed in the execution of the aid-financed projects according to schedule was often considered as an indicator of self-help and of the effective use of foreign assistance.

The pattern of sectoral allocations which emerged from the model were significantly different from that finally adopted in the Five-Year Plan in only three sectors—those of physical planning and housing, transport and communication, and electrical power generation. The sector of physical planning and housing was provided with a smaller proportionate allocation in the Plan than that given in the model—this did not jeopardise intersectoral consistency, since other sectors did not rely to any significant extent on its output; residential construction and urban development could be slowed down without immediate adverse affects on the directly productive sectors. In contrast to housing, the allocation to the two latter sectors was increased to provide for major replacement expenditures; there was a large backlog of urgent repairs and replacement in the transport and communication sector and there was a large number of ongoing power generation projects. There was also a need for investment to modernise and balance the existing capital stock with a view to increasing its productivity.

Non-economic considerations also played a part in the inter-sectoral allocations in the Five-Year Plan. In the first instance there were political commitments to a few specific projects and programmes such as the Nuclear Power project, a Bridge Project linking the northern and southern regions of Bangladesh, and a Coal Mining Project, all of which were considered high priority projects for Bangladesh before independence. They figured prominently in the manifesto and election campaigns of the ruling political party. A few of these projects had been matters of great tension before independence between the Pakistan government and the then provincial government of East Pakistan. The Pakistan government had not been enthusiastic about them and had delayed their consideration and/or implementation. Such delays were considered intentional and as acts of discrimination against East Pakistan. Some of these projects required reappraisal in the light of changed circumstances after independence, including changes in costs and market prospects. But they could not be abandoned or postponed by the political authorities without a considerable loss of face. Similarly, large-scale flood control projects were considered crucial in view of the frequent floods in Bangladesh which resulted in considerable loss of life, income and assets every year. There was, therefore, great pressure to invest in flood control projects, many of which had been under examination and scrutiny for several years before independence. There were differences of opinion between the Planning Commission and the administrative ministries about the state of preparedness of a few of the projects; the Planning Commission wanted new and more careful studies. A few of these projects were tentatively included in the Plan with the explicit understanding that further studies would be undertaken by the ministries before projects were finally approved.

Whenever there was a great deal of uncertainty about future demand and supply conditions and whenever the knowledge of alternative investment possibilities was very limited, there was a bias in the Plan against making commitments extending over long periods. This was particularly true in the case of projects which yielded output over a very long period. Whenever the area of uncertainty was large, the investment package was kept to the minimum so that it was possible to start with a small programme and build up to a larger programme as facts and analysis accumulated. This was especially so in the case of investment projects in the transportation and power sectors. New studies relating to the investment possibilities in these two sectors were initiated at the same time as the Five-Year Plan was being formulated. In both the sectors investment decisions were kept down to a hard core of projects, which were intended to fill in the obvious gaps, generally in the nature of replacement, modernisation and balancing of the existing capital stock. The investments were to be made, as far as possible, in such a way that enough flexibility would be retained to make adjustments without significant costs being incurred in the future as and when the results of the studies under way were available.

A considerable amount of flexibility was also adopted in the health and family planning programmes; there were many controversial and unresolved issues relating to the appropriate institutional framework for and technology of the population control programme. Constant research and evaluation of the existing programmes were considered essential in view of the highly unsatisfactory state of knowledge about the causes and motivations for human fertility behaviour and also in view of the recognised shortcomings of the past and existing programmes. Thus it was sought to build a large component of experimentation into the Plan, both in respect of alternative techniques and of institutions to be introduced for the population planning programme.

The Five-Year Plan did not include regional aspects of development explicitly in the input–output model. It did not split up the overall development programme into regional or district components, nor did it work out the implications of the Plan for the growth of income and employment in the different regions. Part of the reason for this was the high degree of mobility of labour within a compact, homogeneous and geographically small country like Bangladesh. It was assumed that the location of development projects would be determined by technical and economic considerations. However, the regional aspect of development was politically important, especially with respect to the expansion of employment opportunities. An indication of its significance was evident from the government policy of imposing regional or district quotas for employment in the public sector agencies.

There were a few projects, however, which were promoted with the explicit objective of promoting regional development. The most significant project in this respect was the construction of a bridge over the River Jamuna (Brahmaputra), linking the northern and southern regions of Bangladesh. This was a mighty river, carrying large quantities of water during the rainy season and frequently causing floods. Furthermore, it often changed course. Both these features presented difficult engineering and technical problems and involved very expensive structures. The project was considered not only to be very expensive and difficult to implement but also had a very long gestation lag. Against this, it had very important regional aspects since it was expected to provide a very effective link between the two regions. The northern region had been relatively underdeveloped compared to the south. The bridge was considered important for the development of the northern region and to serve as a mechanism for the diffusion of the effects of development in the south towards the north, by increasing interregional trade and factor movements. Similarly, the generation and transmission of power facilities had an important regional aspect. The eastern region had a surplus generating capacity as compared to its requirements whereas the western region had inadequate generating capacity. There were a number of choices, each of which had different regional implications. The two regions, for example, could be independent in terms of their power supply so that generation capacity in each region could be developed to meet its present and future requirements. Alternatively, the eastern region could supply power to the western region through an interconnecting power grid. Again, instead of generating electricity in the eastern region and transmitting it to the western region, gas could be supplied to the western region by pipeline and generating capacity could be built up on gas in the western region.

The development programme, especially the agricultural production programme. was expected by its very nature to have a different impact on each of the geographical regions. In the first place, the rice production programme was concentrated in the "flood-free" areas, which experienced little flooding during the monsoons; moreover, within the flood-free areas, the spread of new technology, including the use of new varieties of seeds and fertilizer would be uneven, depending not only upon the speed of institutional development but also upon the types of irrigation facilities available in different areas. The irrigation programme was unevenly distributed because of the dependence of pumps and tubewells on the availability of adequate water supplies. The deep tubewell programme was governed by the availability of ground water and was concentrated in the north-eastern region whereas the low-lift pump irrigation programme, which was dependent on surface water was more widely spread. Each type of irrigation, therefore, had its own growth potential in different regions and had to proceed at different speeds. The uneven regional impact of the agricultural production programmes could only be counteracted by a policy of geographical dispersion of the rural public works in such a way as to promote regional equity.

(e) Salient Features of the Five-Year Plan

The foregoing analysis illustrates the wide range of considerations and factors, partly aided by the use of a consistency model and partly modified by the judgement of the planners, administrators and politicians, which entered into the formulation of the size, priorities and strategy of the Five-Year Plan.

The Plan provided for a financial outlay including non-investment development expenditures of about Tk.44,500 million or about US $5.57 billion in 1972/73 prices. The size of the Five-Year Plan was closest to one of the alternatives included in the planning model, which was referred to as the medium Plan and provided for a 2.5% rate of growth in annual consumption. The gross financial outlay as a percentage of GDP was expected to be about 12% in 1973/74 and to increase to about 19% by 1977/78 whereas gross domestic financial saving was expected to reach 4.5% of GDP in 1973/74 and to increase to about 14.2% by 1977/78. The foreign capital inflow was expected to finance about 40% of the total development expenditures for the entire Plan period. The Plan was intended to achieve a 5.5% rate of growth of GDP per annum with a per capita rate of growth at 2.5%. The rate of growth of population was assured to decline from 3% per annum to 2.8% by the end of the Plan period. Per capita consumption was expected to increase by 18% over the Plan period. Among the principal items of consumption, rice consumption per capita was expected to be no more than 1.2% higher, whereas textile consumption was expected to be no more than 8.5% higher.

The Plan intended to eliminate the import of foodgrains by the end of the five-year period and to provide for an expansion of employment opportunities sufficient to absorb the increase in labour force. No major structural change in the composition of GDP was expected; agriculture would continue to be the predominant component of GDP, declining from 57.6% of GDP to about 55.1% between the benchwork period and 1977/78; the share of the manufacturing sector was to increase from 10.4% to 11.2% of GDP, the share of services remaining more or less constant.

Agricultural and water resources were allocated 24% of planned development expenditure. The high share of industry, 20% of the total, was mainly due to a high rate of import substitution provided for in the three important industries of cotton textiles, cement and petro-chemicals. It was hoped to introduce domestic production of many of the items of industrial production imported from Pakistan before independence. The petro-chemical complex, based on the natural gas in Bangladesh, was seen not only as an import substitution industry but also as an eventual export industry, particularly of fertilizer. Expansion in the domestic supply of cement was seen as a critically important component of investment expenditure in Bangladesh in order to ensure that it would be readily available during the construction season in the winter months. The investment programme in the transport and communication sector (20% of the total) had a large component for balancing and modernisation; the programme was to be kept flexible until the results of a transportation survey then under way were available. The emphasis in the power development programme was on the transmission and distribution of electricity. This combined with the programme of development of natural resources constituted 2% of total investments. Compared to the size of their development programmes in the previous plans, the social sectors received a proportionately higher allocation than ever before to compensate for past neglect.

(f) The Process of Approval of the Five-Year Plan: the Role of Political Leadership and Public Opinion

The Five-Year Plan, as outlined above, and after it was prepared by the Planning Commission, had to go through a process of approval by the political leadership and the Cabinet before it was accepted as a basis for the determination of the detailed programmes and policies. What part did the political leadership play in its deliberations and in its approval or modifications? Was there a public debate on the Plan or any attempt to elicit public opinion on the recommendations of the Plan?

The broad outlines of the Plan were discussed by the Planning Commission in general terms with the Prime Minister who became the Chairman of the Planning Commission in early 1973. The Prime Minister approved the broad outlines of the Plan without any detailed examination of the implications of all its policy recommendations. He expressed great concern over the state of mass poverty, unemployment and shortage of food and wanted the Plan to be designed to provide for the maximum expansion of employment and food production. He agreed with the Plan's emphasis on agricultural growth as the key to the economic progress of Bangladesh. In general terms, he emphasised the importance of small-scale, cottage and household industries, as an important vehicle for generating employment in rural and semi-urban Bangladesh.

The Planning Commission did not prepare a preliminary "think-piece" or approach paper stating the critical policy issues and choice regarding size, resources, objectives and priorities of the Five-Year Plan for consideration by the Cabinet, National Economic Council or by the Parliament. This was partly due to limited time; partly it was felt that it would facilitate more constructive comments and fruitful discussions if the entire Five-Year Plan was prepared in draft form before consideration by the Cabinet and, it was hoped, by Parliament as well. In the event, Parliament did not debate the Plan. The parliamentarians had the responsibility of enacting laws which would implement the changes in policies and institutions implied in the Plan. It was essential that the executive branch of the government and the planners should understand how the politicians would have reacted to their policy suggestions and how easy or difficult it would have been to carry them out. A debate by the Parliament was ruled out on the legalistic ground that there was no provision in the Constitution for parliamentary approval of the Five-Year Plan. Since the Five-Year Plan was a guideline and a framework for taking a decision on specific projects or policies, rather than an operational document, it was argued that what was really important for Parliament was to examine concrete proposals for expenditures, resource mobilisation and economic policies. This Parliament was expected to undertake at the time of the approval of the annual Plans and budgets; it was the constitutional right of Parliament to approve expenditures and tax policies included in the Annual Plan.

It was true that the Annual Plans, which the Parliament debated and approved, were the instrument for the implementation of the Five-Year Plan; however, an appreciation and evaluation of the basic macro-economic framework as well as of the overall strategy of the Five-Year Plan was necessary in order to provide the perspective and background for a critical examination of the specific policy proposals of the Annual Plans. The lack of interest of the top political leadership in having the Parliament debate and discuss the Five-Year Plan was partly a reflection of the limited role which the Cabinet was prone to relegate to the Parliament. However, after the Plan was finally approved by the Cabinet, members of the Parliament were supplied with copies of the Five-Year Plan, not for debate but for perusal.

The detailed discussion by the Cabinet which took place subsequently revealed that most of the comments were concerned with the administrative structure and executive responsibility for the development projects within their respective sectors. The Cabinet did not discuss the macro-economic aggregates of the Plan nor its strategy and priorities. The discussions were mostly confined to the sectoral programmes and that too was restricted to a few specific issues of interest to a few individual ministries. Let us quote a few examples. The suggestion in the Plan relating to the agricultural sector that the ceiling on land ownership should be reduced from 33 acres per family to 8 acres per family was rejected. Again, the Plan contained strong criticisms of the poor performance of the nationalised industries and the tardiness of the government in adopting ameliorative measures. This criticism was considerably toned down as a result of discussions in the Cabinet. Thirdly, the Plan emphasised population planning as distinguished from family planning in the narrow sense of birth control measures and suggested the creation of a new administrative unit to be called the Population Planning Division, to co-ordinate the entire range of educational and motivational activities undertaken by various development ministries. In the face of strong opposition from the Ministry of Health and Family Planning, i.e. that this would detract from the importance attached to the family planning programme centred around rural health centres, the Cabinet accepted the general concept of a Population Planning Division but deferred for a future decision the question of the location and functions of the Division.

Yet another example related to the question of an appropriate implementation agency for the excavation of derelict tanks and the development of local markets and market towns. The Plan with its emphasis on decentralisation suggested that these projects should be the responsibility of local government institutions, which should contribute towards their financing by mobilising local resources, including the mobilisation of the unutilised rural manpower. This suggestion was opposed by the ministry which in conformity with traditional practice and in the interest of speedy implementation, wanted to administer the projects directly with the help of contractors, centrally appointed and controlled. The final decision was to keep the choice between the two alternative methods of implementation open for future decision. The preoccupation of the ministries with widening their jurisdiction and enlarging their executive powers and responsibilities was also reflected in the ministry of cooperation pressing for a larger allocation of resources for the development of industrial cooperatives. Similarly, the Agricultural Ministry argued against the division of responsibility for the development of irrigation facilities or for the credit and extension facilities between the several ministries, and wanted the integration of all services under its control.

It appears from the above that the participation by the Cabinet in the examination of the Five-Year Plan was very limited, haphazard and disjointed. The overall perspective of the Plan was not subjected to serious examination. Each ministry concentrated not only on its own sector programme but also on those aspects of the programme which related to it administrative responsibilities and its relationship with other ministries. It was the institutional aspect which drew the major—almost exclusive—attention.

Several suggestions were made by the Planning Commission about the various ways in which public opinion should be elicited in the course of the formulation of the Plan, especially in generating collective thinking on and participation in the making of the Plan. It was proposed that the draft Plan should be published and thrown open to public discussion. Newspapers and other information media would be expected to elicit and express criticisms and comments on the Plan. Various professional groups or associations (such as those of economists, engineers, doctors, and other professions, etc.) were to be invited to hold symposia or conferences to discuss the Plan.

The Planning Commission also proposed to hold a conference of economists, statisticians and allied professions to evaluate the statistical data base of the Plan as well as to examine the macro-economic aggregates postulated in the Five-Year Plan.

The suggestions noted above did not find favour with the top political leadership.[5] It was contended that the various suggestions made by the Planning Commission for eliciting comments on the draft Plan would, in fact, be tantamount to a public debate, possibly raising unnecessary controversies or peripheral issues without throwing any light on the substantive matters. It was argued that the experience in 1972 with the process of constitution-making, which attempted to elicit public opinion and comments, was disappointing. It did not contribute much, in the opinion of the political leadership, to the process of constitution-making. In fact, it was argued that the public awareness or knowledge of economic issues would be very limited to allow a very intelligent discussion of the Five-Year Plan. This line of reasoning misunderstood the purpose of wide public participation in the discussion on the Five-Year Plan. The purpose was not so much that it would contribute new ideas and insights but that it would educate public opinion, various interest groups and political parties about the objectives of the Plan as well as about the policies and institutions which were required for its implementation. It was necessary to promote a widespread realisation of how the various interest groups and the distribution of economic power would be affected by the Plan. Participation by the various interest groups in the debates and discussions on the Plan, even if as a result no change was made in the Plan, was helpful in generating support and enthusiasm for it amongst the public. A Plan was a socio-political document and its success was facilitated by providing a wide sense of participation in its formulation.

Notes

1. *Address of the President of the Peoples' Republic of Bangladesh to Parliament*, June 1973. Ministry of Information, Government of Bangladesh.

2. World Bank, *Land and Water Resources Sector Study—Bangladesh*, Vol. III, Washington, December 1972.

3. Other exogenous investment requirements were used for sugar, paper, generation of power, gas pipeline, flood control, health, education and family planning.

4. (1) There. was the big plan with a GNP growth rate of 6.5% per annum and a 3% annual rate of growth of per capita consumption.
(2) The alternative of this plan laid down a higher level of exports.
(3) The medium Plan postulated a 2.5% rate of growth of annual

consumption, the same rate of growth of exports as in (1), but with a 20% reduction in exogenous investments below the level in (1), except for the family planning programme.

(4) The small Plan postulated a 2% annual rate of growth of consumption, the same rate of growth of exports as in (2) and a 25% reduction in exogenous investments, with the exception of family planning which was reduced by only 10%.

(5) The very small Plan postulated a 1.5% annual rate of growth of consumption, the same rate of growth of exports as in (1), and a 30% reduction in exogenous investment except for family planning which was reduced by 15%.

5. The Planning Commission in Bangladesh did not hold a conference of economists. It was unlikely that the political leadership would have prevented the Commission from formally consulting academic economists or from constituting a panel of economists. The Bangladesh Economic Association, at its annual conference in 1974, mainly discussed the Five-Year Plan, which had already been in operation for a year. They were more critical than usual, partly because they had been excluded from participation at the earlier stage. This exclusion seemed to them an indication of the intellectual arrogance of the planners, as well as their hyper-sensitivity to criticism.

5A. The Annual Plan as an Instrument of Planning

(a) Introduction

The Annual Plan was an integral part of the planning process; it was an instrument for the implementation of the Five-Year Plan through a detailed formulation of projects, programmes and policies from year to year. The Five-Year Plan attempted to indicate very broadly how the key variables delineated in it were likely to evolve over time; this was a very tentative exercise and was intended to illustrate in quantitative terms how the efforts needed to carry out the Plan would have to be built up, if targets and objectives were to be fulfilled. The picture that emerged for some principal components of the plan is shown opposite.

ANNUAL PHASING OF FIVE-YEAR PLAN[a] IN 1972/73 PRICES
(million Takas)

	Development Expenditures			*Imports*		*Exports*		
	Public	*Private*	*Total*	*Food Imports*	*Non-food Imports*	*Total (after adjusting for terms of trade loss)*	*Visible and Invisible Exports*	*Loss in Terms of Trade*
1973/74	5,250	700	5,950	—	5,400	3,400	3,400	—
1974/75	6,450	800	7,250	—	6,610	3,940	4,020	–80
1975/76	7,750	950	8,700	350	7,290	4,280	4,410	–130
1976/77	9,250	1,150	10,400	—	8,020	4,640	4,830	–190
1977/78	10,820	1,430	12,250	—	7,010	5,030	5,300	–270
	39,520	5,030	44,550		53,780[b]	21,390	21,960	–570

a. *The Five-Year Plan*, ibid., pp. 35, 52, 58, 64.
b. Including (*a*) Tk.100 for delays and uncertainty and (*b*) unallocated invisible payments of Tk.450 million.

No attempt was made to calculate in detail how domestic savings were expected to develop although a tentative or illustrative phasing of the additional tax receipts over the Plan period was undertaken.[1] This is shown opposite.

Great difficulties were encountered in an attempt to depict the flow of resources and expenditures over a five-year period. The phasing of the value of foodgrains imports, for example, raised

YIELD FROM ADDITIONAL TAXES

Year	*(Million Takas)*
1973/74	250
1974/75	600
1975/76	1,100
1976/77	1,750
1977/78	2,550

special problems because it depended not only on fluctuations in domestic food output but also on international price movements. It was not possible to phase certain items of invisible payments such as those for shipping. Moreover, it was necessary to make provisions for uncertainties and delays in the execution of plans and programmes. The phasing of the decline in the terms of trade was made on the basis of a heroic assumption and much significance could not be attached to it. It was the task of the Annual Plans to undertake and implement the phasing of the Five-Year Plan.

Secondly, the Five-Year Plan provided mainly sectoral and sub-sectoral allocations, except where important, very large projects were involved. The Annual Plans were, therefore, intended to formulate the individual projects in detail and to undertake their detailed phasing over time, in terms of both financial expenditures and physical progress. Ideally the macro-economic framework of the Five-Year Plan should have been formulated simultaneously with the sectoral programmes and projects, because they were interdependent. In practice, all the project studies and detailed analysis of costs and benefits of individual projects were not available early enough to be incorporated into the formulation of the Five-Year Plan. It was left to the Annual Plan to examine and specify the individual projects and to approve the investment alternatives. The Annual Plan aimed to achieve some degree of consistency between the sector and sub-sector programmes. This was not to be carried out by the use of any formal economic model; it rested mainly on inter-ministerial consultations. It was expected that as additional information and results of new studies or analyses were available during the course of implementation of the Five-Year Plan, the Annual Plans would serve as the mechanism for introducing changes in sectoral allocations, policies and institutions. The adjustment of the Five-Year Plan to changing circumstances was intended to proceed through the formulation of successive Annual Plans.

The annual development programmes, as adumbrated in the

Annual Plan, specified expenditures on items and projects approved for inclusion as development expenditures which formed an integral part of the Annual Budget. The annual development programme was the link between the Annual Plan, as prepared by the Planning Commission, and the Annual Budget, as prepared by the ministry of finance. The Finance Ministry accepted the annual development programme as prepared by the Planning Commission and put it in the accounting framework of its Annual Budget. Since the scrutiny of non-development expenditures was considered the prerogative of the ministry of finance, the preparation of the Annual Plan and the Annual Budget provided the only occasion when the Planning Commission was involved in the examination of non-development expenditures. Because resources available for development expenditures in the public sector consisted of the surplus of total revenues over non-development expenditures, the latter, the Planning Commission argued, required as serious a scrutiny in respect of priorities and cost effectiveness as was required of the development projects. There was a trade-off between non-development and development expenditures.

The Annual Plan also undertook a review of the performance of the economy from year to year; in particular it analysed developments in the fields of foreign trade, fiscal and monetary policies as well as price movements. One of its purposes was to evaluate the progress of implementation of projects and sector programmes in terms of both financial expenditures and physical implementation. It spelled out in greater detail the recommendations of the Five-Year Plan for changes in policies and institutions and evaluated them in terms of actual decisions taken and the rate of implementation.

The Annual Plan provided an occasion for a detailed examination of resources, both domestic and foreign, likely to be available for the implementation of the Plan. Aid commitments were usually made annually, and often were linked with a review of the past performance of the economy, including the pace and pattern of utilisation of past economic assistance. As in the Five-Year Plan, so also in the Annual Plan, resources and expenditures were matched not only in the aggregate but also in terms of their foreign exchange components. Foreign exchange budgeting was an important component of the Annual Plan. In addition, exercises in commodity balances were undertaken. There was an attempt to achieve a balance within a given year between supply and demand for essential items of consumption such as cereals, edible oil, coarse cloth, kerosene, sugar, etc., as well as of critical intermediate inputs such as steel, cement, crude and refined petroleum, fertilizers and pesticides.

(b) Public Sector Resource Budgeting

The Annual Plan had to estimate the resources, both domestic resources and foreign assistance, available each year for the public and private sectors. It was on the projection of resources in the public sector that the Annual Plan laid the greater emphasis, in view of the preponderant role of the public sector in the economy. The estimate of public saving was derived by subtracting non-development expenditures from estimated tax and non-tax revenues.

The estimation of tax and non-tax revenues was an exercise jointly undertaken by the Planning Commission and the Ministry of Finance. It took place in two steps. The first step was to project revenues on the basis of existing rates of taxes and fees. This was related to the growth of the economy and its various sectors which were the major sources of revenue. The growth of import taxes was related to the rate of growth and the value of imports; the growth of excise duties was related to the growth of domestic output and of excisable commodities. The second step was to predict how much additional revenue would be produced by an increase in tax rates. It was the second step in the forecasts of resources which involved hard policy decisions. Both the Planning Commission and the finance ministry were involved in making recommendations about increases in existing taxes or fees and the introduction of additional sources of revenue. The finance ministry always made conservative predictions of revenue, on the basis of existing rates as well as on the basis of new rates and additional tax measures. A compromise between its estimates of resources and the more optimistic estimates of the Planning Commission had to be reached. The Planning Commission had an interest in increasing the rate of mobilisation of public sector resources so that a larger public development expenditures programme could be undertaken. The non-development expenditures of the government appeared as a relatively inflexible item. In fact, there was a need for a certain amount of expansion in non-development expenditures arising from the need to expand what was previously a regional administration to meet the requirements of a national government. In addition, there was the need to provide for the increasing cost of administrative services in response to rising prices and wages.

The projection of revenues in the public sector and the growth of the economy were interdependent. The growth of the economy was related to the rate of investment or development expenditures, which in turn were related to the budgetary surplus as determined by the excess of revenues over non-development expenditures. This was the interrelationship which had to be brought to the attention of the political leadership and Cabinet every year.

On the basis of the projected budgetary surplus, a development programme was formulated by the Planning Commission in consultation with the administrative ministries. Decisions had to be taken as to which of the new projects would be initiated and how much of the projected expenditures on the existing projects would be undertaken during a given year. With the passage of time, the costs of the projects in current prices had to be upgraded; throughout the period prices and costs were steadily rising. The "cost overrun" was the usual characteristic of most of the development projects. On the average 20–30% cost overrun was not unusual. This was only partly due to a rise in prices of inputs; the fact that all the necessary information on costs was not available at the time of the preparation of projects coupled with an eagerness on the part of the agencies and ministries to make a project look better, i.e. yield a higher benefit–cost ratio at the time of approval of the projects, also introduced an element of under-estimation. There were two constraints on the choice of projects for inclusion in the annual development programme. One was the need to accommodate the ongoing projects, which was analysed earlier in connection with the discussion on the preparation of Five-Year Plan. The other was the availability of foreign economic assistance in the form of project aid. Owing to a scarcity of foreign exchange most of the projects had their foreign exchange component financed by project assistance. The availability of aid already negotiated or committed for a project would favour the inclusion of that project in the Annual Plan. The pace of negotiation of project assistance and its composition had a predominant influence on the magnitude and composition of the development programme. The donors could determine the pattern of development expenditures by the exercise of their choice amongst the list of projects presented to them for financing. Even when the donors selected from within the list of projects presented by government, they could affect significantly the composition of the development programme in a given year by delaying commitment for some projects and accelerating commitment of funds for others. Thus they could change the priority of individual projects over time within the list of projects agreed upon by the government.

The preparation of development programmes for the Annual Plan required data on actual expenditures incurred in the past year on a particular project. This was necessary in order to be able to decide how much of the remainder of the expenditures on a given project was to be incurred next year and in the succeeding years. Partly past expenditures were a reflection of the absorptive capacity of the administrative ministries. The record of the past expenditures determined the need for as well as the ability to undertake additional expenditures in future years. The data requirements for the Annual Plan were in a sense more stringent than those for the

Five-Year Plan because it needed more up-to-date data. Mostly data on an annual or quarterly basis was required to make projections and estimates for the Annual Plan. The fact that there was a time lag in the generation of data was a severe handicap. The data on expenditures undertaken in the last year were hard to come by, partly because of the accounting and administrative procedures followed by the Ministry of Finance. The allocation of funds by the Ministry of Finance was made in two different ways. The administrative ministries and agencies, as distinguished from autonomous agencies and public enterprises, received an authorisation to spend during the year whatever amount was allocated to them in the Annual budget without reference to the finance ministry. These allocations were made at the beginning of the financial year. There was no way for the Ministry of Finance to obtain details of their expenditures until they were reported by the Auditor-General's office; there was always a time lag in the reporting of such expenditures. Autonomous corporations, on the other hand, could incur expenditures only after obtaining specific quarterly releases from the Ministry of Finance, even though the expenditures were provided for in the Annual Plan. In this case, the quarterly allocations made by the Ministry of Finance were recorded in their books, although the expenditures actually undertaken could only be known after a time-lag of at least six months. Sometimes, expenditures were over-estimated, because the autonomous agencies would draw upon the allocations and would keep them in their independent bank accounts, carrying them over to the next year. The Ministry of Finance released funds only on a quarterly basis in order to keep a rough balance between inflow and outflow of funds; since the amount of quarterly releases of funds by the ministry of finance was uncertain and often delayed due to the overall financial stringency, the agencies were unsure as to when and how much of the allocations they were likely to receive so that they kept a reserve balance for meeting their ongoing expenditures.

Incomplete information on the flows of financial expenditures was matched by even scantier information in respect of the physical implementation of projects; it was difficult to be sure about the correspondence between financial expenditures, on the one hand, and progress in physical implementation of projects, on the other. The existing information system on the physical implementation of projects was not only fragmentary but also took a long time to be collected. A revised system for collecting information was introduced in 1974/75 and was intended to facilitate an evaluation of the physical implementation of projects. The information was required to be presented in some detail.[2] Information on progress in physical implementation was also sought and it included details of construction of such categories as buildings, roads, bridges, drains and embankments, etc., with specifications of the physical measurements in each case. Similarly, progress in the installation of various pieces of machinery and in the procurement of transport equipment was to be specifically indicated. In spite of a revised system of gathering information, the agencies and ministries lagged behind in the supply of information, partly because of inadequate manpower resources engaged in this task.

(c) Foreign Exchange Budgeting

Foreign exchange was a critical constraint in the economy of Bangladesh and an elaborate mechanism, using both price and direct controls, was built to allocate foreign exchange resources between alternative uses. The degree of control fluctuated with the degree of stringency of foreign exchange. The greater the stringency of foreign exchange availability, the greater was the element of control. At the time of the preparation of the Annual Plan, a foreign exchange budget was worked out, indicating in detail the sources and uses of foreign exchange resources. The foreign exchange requirements were worked out in considerable detail both for items of imports and in terms of users, private and public; in the case of the public sector users, the different agencies or ministries were specified. The requirements of imports of capital goods were directly related to the foreign exchange components of capital expenditures on development projects. The import requirements of intermediate inputs such as raw cotton, cotton yarn, oil, cement and chemicals, etc,. were related to the utilisation of capacity. The essential consumer goods like foodgrains, edible oil and sugar, etc., were related to the need to prevent a fall or even to allow a certain increase in per capita consumption of essential commodities.

Foreign exchange budgeting was complicated by the fact that sources of financing were many and various. There were Bangladesh's own foreign exchange earnings, commodity or non-project assistance (often specified in terms of commodities), and project assistance which was tied to the financing of the foreign exchange component of the aided projects. There was a limited amount of project aid which financed the local cost of projects thus supplying free foreign exchange for balance-of-payments support. Barter trade, notably with the centrally-planned economies, tied the procurement of imports to exports from Bangladesh. Thus exchange earnings from exports under barter agreements did not constitute "free" foreign exchange, available to finance any part of the import bill. Thus, the different sources of foreign exchange were not substitutable, since they were designed for specific purposes. This meant that foreign exchange available for particular commodities or uses had to be matched with the import requirements for those items or uses. Given the total amount of commodity and project assistance available in a particular year, the objective was to assign as many items and as much as possible of imports of consumer and intermediate goods to commodity assistance and as much as possible of capital goods imports to project assistance. This allowed Bangladesh's own foreign exchange earnings or "free" foreign exchange to be used with greater flexibility in order to meet unforeseen demands or shortfalls in supply. Similarly, the utmost effort was made to utilise exchange earnings from exports under barter agreements in order to avoid accumulating unspent foreign balances.

The claims on untied foreign exchange resources which received the highest priority were external debt service payments and invisible payments—none of which could be financed by commodity assistance. The next most important claims on free foreign exchange were for imports of food and crude and refined petroleum; in spite of food aid, Bangladesh had to buy substantial quantities of food. The rest of the foreign exchange earnings was very largely utilised for imports of other consumer and intermediate goods, leaving a small margin for capital goods. This component of the foreign exchange budget was administered by the Ministry of Foreign Trade which licensed imports to both public and private sector importers. Imports of capital goods were predominantly financed by project assistance; only a small part of foreign exchange earnings (no more than 5%) was used for the import of capital goods. Non-development imports for the various ministries and public sector agencies (117 in number), excluding the public sector productive enterprises, constituted another 5% of foreign exchange earnings.

This brings us to the administrative procedures for the allocation of foreign exchange resources under the foreign exchange budget. The Planning Commission and Ministry of Finance played the key role in the preparation of the foreign exchange budget and in making broad sectoral allocations. They jointly estimated the foreign exchange resources available from the different sources, in consultation with the Ministry of Foreign Trade. A committee composed of the Planning Commission and Finance Ministry allocated resources between broad categories while another committee representing the Ministries of Finance, Foreign Trade, Industries and the Central Bank formulated the detailed foreign exchange requirements under each of the broad headings. In addition to carrying out their task of allocating foreign exchange resources in conformity with the priorities of the Annual Plan, the committees acted as a watchdog on the flow and rate of utilisation of foreign exchange, reviewing the position every quarter or every six months. Bangladesh operated the foreign exchange budget on a very narrow margin of reserves. It was always uncertain how much foreign exchange resources would be available because there might be shortfalls in exchange earnings, delays in the receipt of foreign economic assistance and lags in the utilisation of foreign exchange resources. The level of foreign exchange reserves had to be watched constantly.

Timely information on the flow and rate of utilisation of foreign exchange resources was difficult to come by and this complicated the task of foreign exchange budgeting. The rate at which import licences were utilised by the private and public sector agencies was known only after some time elapsed and full information on it was never available at the time when the six-monthly allocations of foreign exchange were revised towards the end of the licensing period in the light of events. There was a time lag between the issue of licences and the opening of letters of credit as well as a lag between the opening and the retirement of letters of credit. Only the last step was reflected in the state of the foreign exchange reserves as revealed at any point of time. As a result outstanding liabilities for foreign exchange payments were carried over from one six-monthly period to the next. No detailed information on the utilisation of import licences was available because the prevailing accounting and administrative system was not designed to collect and collate information on foreign exchange payments at frequent intervals, monthly or quarterly, so that it was not possible to predict when past commitments would fall due for payment. All these factors made it extremely difficult to control the working of the foreign exchange budgeting system. There were periods of panic when foreign exchange reserves fell drastically as well as periods when there was temporarily, an unintended accumulation of foreign exchange reserves. Consequently, there was a pattern of "stop–go" in import licensing with short periods of liberal licensing being followed by drastic curtailment of foreign exchange expenditures.

(d) Food Budgeting

Food budgeting was an important aspect of the Annual Plan, since minimum availability of food, the basic human need, was essential for political and social stability. More than two-thirds of average consumption expenditures were on food and the proportion was higher in the rural areas and amongst the lower income groups. The deficiency in calories and protein consumption of the majority of the population was a matter of serious concern in view of widespread malnutrition. To ensure the supply of what was considered the minimum essential intake of cereals and of a few non-cereal items of consumption such as edible oil, salt, sugar, kerosene oil and coarse cloth, was accepted as a part of the government's responsibility towards meeting the basic needs of the population.[3]

Starting in the mid-1950s, there had been a continuous rise in the imports of foodgrains as a result of the rapid growth of population coupled with a stagnant or slowly rising output of food. Programming imports of foodgrains, as well as of edible oil, sugar, kerosene and other commodities in advance for the year as a whole, was an important part of import policy and foreign exchange budgeting since domestic production of all these items fell far short of requirements. Programming supplies of foodgrains was complicated by fluctuations in the domestic output of foodgrains. Fluctuations in the marketable surplus were considerably greater than fluctuations in output, for only production in excess of growers' requirements was placed on the market and even in good years this constituted only a small part of domestic output. To make matters worse, hoarding and speculation by farmers and traders added to fluctuations in the availability of foodgrains on the open market.

Programming of food supplies involved the government in a number of difficult choices. It was necessary to choose between importing foodgrains at the beginning of a year in order to build up stocks at home against an unexpected shortfall or earmarking foreign exchange reserves for the purchase of food in case it was needed during the year. Since food budgeting was done at the beginning of the year, and since estimates of the food deficit were always uncertain, holding a certain amount of foreign exchange resources in reserve to import foodgrains, as and when needed, saved the cost of storing foodgrains, particularly as the limited storage facilities in the country meant that supplies might deteriorate or be lost. On the other hand, there was considerable uncertainty about the availability of foodgrains in the world market for purchase at times of need.

Imports had also to be geared to seasonal variations in the amount of a good needing to be distributed through the rationing system. This was because the amount of rice put on the market fluctuated between the different seasons. The farmers often sold at the time of harvesting and bought back a part of their requirements later in the season, in order to match their need for cash at different times of the year. At the time of harvesting a smaller amount of foodgrains was distributed under the rationing system than at other times. Shortages were most acute when farmers' and traders' stocks held from the previous crop were exhausted before the next crop was harvested. It was then that demands on the public distribution system were at their highest and adequate supplies during these "lean" months had to be ensured by the government.

Fluctuations in the availability and price of foodgrains were a further complication making it difficult to decide when to buy in the world market. In view of the seasonal fluctuations in the price of imported food, there was an advantage in purchasing when prices were low in the world market. In view of the shortage of foreign exchange, on the other hand, it was not always possible to make purchases of foodgrains advantageously when prices were low in the world market. The inflow and outflow of foreign exchange resources had to be matched from one six- or three-month period to the next. Added to this was the need to arrange for foreign shipping and transportation in time as Bangladesh had no merchant fleet which could be directed to transport emergency supplies of grains.

Still another complication arose because of Bangladesh's heavy dependence on food aid. Commitment of the entire amount of available food aid that would eventually be forthcoming was not made at the beginning of the year. Negotiations continued throughout the year so that it was difficult to ensure that predetermined amounts of food aid would be received by Bangladesh in various seasons or months to match the varying domestic requirements. Uncertainties also arose in relation to the shipment of food aid. In the case of some countries providing food aid, Bangladesh was not free to ship food in vessels of her own choice, since food aid had to be transported in the vessels of the donor country. The irregular nature of commitments and disbursements of food aid resulted at times in bunching together shipments of food aid and domestic purchases of grains, quite unmatched to the seasonal patterns of food marketing and distribution in the country. Over the years an attempt was made to obtain an assurance, if not a firm commitment, from the donors about the amount of total food aid to be made available in a particular year, particularly after the formation of the Aid to Bangladesh Consortium in 1974/75. In the earlier years, Bangladesh had to plan her food budget on her own assessment of expected food aid, on the basis of preliminary consultations with the donors, and provide for her own purchases on the basis of an uncertain expectation of food aid. There was a considerable lag and lead in the arrivals of food imports on both counts with sometimes dire results.

As has already been pointed out, food budgeting was basically directed towards meeting the requirements of the public distribution system. In this respect there was a choice between importing food and procuring it domestically for distribution through the rationing system. Even when there was a food deficit for the country as a whole, the surplus farmers (those who had a surplus above their consumption requirements) consumed more food per head than the national average; moreover, they increased their consumption immediately after the harvest. Among those who did not produce food in the urban and the rural areas, the higher income groups consumed more food than the national average. The rationing system was intended to supply food to consumers in the urban areas and those in the rural areas who were too poor to meet their needs at the ruling market price. To the extent that the government was able to procure food from the surplus farmers, it would be possible to reduce their excess consumption and to import less, It was also thought that purchases of food in the domestic market might reduce the stocks of food available to farmers and traders for speculative operations at times of shortage. The domestic procurement could be used to help the government meet its obligations under the public distribution system when imports were inadequate. Procurement from domestic production could also be seen as a means to build up government stocks to meet the impact of national disasters or unforeseen shortfalls in domestic output.

Domestic procurement and rationing of foodgrains involved decisions about prices. The purchase price had to be remunerative so that it did not discourage production. This conclusion was true whether procurement was compulsory or made through the open market since in practice only surplus production was procured. There was the possibility that voluntary purchase, even if the purchase price was not unfavourable compared to the ruling market price, might not secure enough foodgrains for the government, if farmers or traders decided to hold on to their stocks in anticipation of future shortages.

The techniques of food budgeting attempted to take account of the considerations outlined above. An attempt was made at the beginning of the year to forecast the domestic output of the three rice crops and the one of wheat. Wheat output was very small; it was less than a million tons while the output of rice varied between 10 million tons and 12 million tons. The estimation of rice output for the purposes of food budgeting were partly based upon forecasts of the Ministry of Agriculture. The statistical procedures for estimating rice output left much to be desired, even though they were improved over time. The estimate was partly based upon forecasts of the use of fertilizer, high-yielding seeds, and water for irrigation. These inputs were critical in determining the output in areas which were using modern seed-fertilizer technology.

The Planning Commission co-ordinated the inter-ministerial discussions relating to the determination of the forecasts of annual food output. The ministries involved in making estimates of domestic output were the Food Ministry, which was in charge of the rationing system, and the Ministry of Agriculture, which was in charge of the production programme, including the procurement and distribution of agricultural inputs, as well as credit and extension services. The Ministry of Agriculture tended to lean towards high estimates, since as a ministry in charge of increasing agricultural production, a high output appeared to reflect favourably on its performance; the Food Ministry had a bias towards a lower estimate, because if there was a shortfall in output, there would be pressure to distribute more food through the public distribution system, which it would not be able to meet. It was in favour of building up a comfortable stock of foodgrains through imports to allow its public distribution system to function smoothly. The ministry also knew that when procurement was counted on to provide part of the supplies to be distributed, failure to reach the target would reduce the amount estimated as procurable, leaving the ministry with a shortfall in supplies available for the public distribution system. The Planning Commission usually found itself in the middle ground in its analysis of future productivity increases.

The estimate of food outputs was revised periodically within the year in response to additional information on sowings, on weather conditions and rainfall as well as on the distribution and use of inputs. The estimates were also revised at the time when the three rice crops were harvested. At the end of the month of June, when the first estimate of output for the preparation of the Annual Plan was made, the Aus crop had already started coming to the market. This was the rice crop which was harvested during the monsoon season. The Aman crop (harvested in November/December) which provided almost two-thirds of total food output was the most crucial one for food budgeting. The final estimate for it was available only in the middle of the financial year. This coincided with the half-yearly review of the Annual Plan as well as of the foreign exchange budget.

There was one factor which introduced an optimistic bias in the forecasts of production of foodgrains. Bangladesh relied heavily on foreign economic assistance for financing her food imports. The aid donors expected Bangladesh to improve her performance in the field of food production and, therefore. to reduce her dependence on food aid. Therefore, while submitting her requests for food aid to the various donor countries, which was done at the beginning of the financial year when the Annual Plan was prepared, Bangladesh felt the need to present a favourable estimate of food output and a diminished need for food aid. Subsequent forecasts of the major (Aman) crop harvested might show a different picture. As years went by, targets for production became less optimistic. There was an increasing realisation of limitations on the implementation of the agricultural development programme. There was also a more realistic appraisal of the expected yield from the application of modern technology in agriculture.

In order to arrive at an estimate of the net availability of foodgrains for domestic consumption, a deduction was made at the rate of 10% to make allowance for wastage and seeds. This was a rough estimate of loss and wastage and was considered low compared to the experience of other similarly situated rice-producing countries in the region, A further step in the preparation of the food budget was the estimation of domestic consumption. This was highly dependent on the estimate of the total population for the year and of average per capita consumption (generally assumed to be 15.5 ounces per day per capita). The difference between estimated domestic output and consumption gave the estimated food deficit. There was no way of estimating either private stocks or marketable surpluses. The latest price trends of foodgrains provided some indirect check on supply, demand and market transactions in foodgrains, but they played no part in determining the estimate of food deficit or import requirements.

Bangladesh's first census of population took place in 1974. The estimates of population which were used until then, were based on projections of the last census held a decade ago, supplemented by various surveys. In 1974/75 the preliminary results of the first census were seen to be widely different from the estimated population figures based on past trends. The preliminary census estimate indicated a figure of 71.3 million for the population against the estimate of 78.6 million used in the Annual Plan for 1974/75. However, the food budget was prepared not on the basis of preliminary census figures but on the basis of higher figures indicated by earlier studies. This created confusion, especially in the minds of the aid donors, because the estimate of the food gap and hence the donors' decision on how much food assistance to give to Bangladesh, was critically dependent upon the estimate of population. The reason for using the estimates previously compiled was given as follows:

"The experience of Population Census in the sub-continent demonstrates that the Census counts usually underestimate the population. This particularly happens for rural and female population. The Census estimate of 1961 for the population of Pakistan had to be adjusted upward on clear evidence provided by demographic research done with the census figures. An under-estimation of 5–8% is considered normal by those who have experience with census. The law and order situation in early 1974 was less than normal and that might also have affected the census enumeration. Taking 7% under enumeration, the population total in January 1974 approximates 77.0 million. On that basis, the estimate of 78.6 million for mid-1974/75 (January 1975) should be considered reasonable."[4]

The difference between domestic requirements and net availability of foodgrains was only one component in the estimate of food deficit or in the determination of import requirements. Some allowances were always made for stocks held by the public distribution system; allowance largely based on guesswork was also made for the changes in private stocks. This was particularly relevant for 1972/73, the first post-war year for which the food budget was prepared. In this year the estimate of the food gap made an allowance for an increase in the consumption of self-grown food by the farmers to compensate for a fall in the level of consumption and hence nutrition in the war years. An allowance was also made for the uncertainties of weather and the failures in the implementation of the programme for the supply of modern agricultural inputs. Target estimates of food production were often considered as ideals to strive at and as a means to keep pressure on the implementing agencies to achieve maximum production. A cushion against a shortfall in the development or production programmes was often provided for in the estimate of the food deficit. The adjustments outlined above were largely matters of judgement and they varied from year to year. In 1972/73 the gap between consumption requirements and the rather optimistic projection of the net domestic availability was estimated at 2.85 million tons, because of the need to rebuild stocks and alleviate wartime privations. It was also feared that the drought of the previous year might be repeated and cause great distress to an already seriously undernourished population. Again during 1974/75, even though the statistical gap between the target of production and the estimated consumption requirements was about 0.61 million tons or 0.86 million tons, including an increase in publicly held stocks of about 0.25 million, import requirements were estimated at 1.7 million tons. Actual *ex post* comparison indicated that projections of import requirements in the Annual Plans were nearer the actual events than would have been the case if import requirements were based merely on the statistical gap between domestic production and consumption requirements.

When the domestic procurement of foodgrains was started in 1973/74 the results likely to be achieved had also to be assessed. The final step in food budgeting was the allocation of Bangladesh's own foreign exchange for the import of foodgrains, and negotiations for food aid to meet the balance of expenditures. The relative importance of the two sources of financing food deficit changed over time. In 1973/74 there was a large increase in the price of foodgrains in the world market requiring additional allocation of foreign exchange to purchase the quantity of foodgrains provided for in the original budget. Divergence between the schedules of commitment and disbursement of food aid also affected the provisions needed from Bangladesh's resources.

Notes

1. *The Five-Year Plan*, ibid., p. 41.

2. The following categories and details were required:

(*a*) original scheduling of work as provided in the project description, over time, year by year, in terms of (1) survey/design, (2) land improvement, (3) construction and (4) installation of equipment and others;

(*b*) original schedule of financial expenditure, year by year, in terms of (1) local expenditure and (2) foreign exchange, of which the foreign aid component was to be shown separately.

3. The government administered a rationing system to distribute foodgrains and a few non-cereal items to all the inhabitants of a few selected urban cities and to selected income groups in the rest of the country.

4. *Planning Commission Memorandum*, "Some Clarification about the Estimate of Foodgrain Production and Import during 1974–75", 6 August 1974.

5B. The Annual Plans and Implementation of the Five-Year Plan

(a) Size and Composition of Annual Plans: Projections and Achievements

The Five-Year Plan of Bangladesh, like all plans everywhere, involved assumptions regarding the working of the economy, and its organisation and management. It contained analysis of the past and the present, predictions about the future and prescriptions about policies, investment programmes and institutions. All plans, their assumptions and predictions, are seldom fully realised. Many events were unpredictable as explained in the preceeding section; forecasts might be optimistic or pessimistic and information was often unreliable or incomplete. The Annual Plan, the basic instrument for the implementation of the recommendations of the Five-Year Plan in concrete details from year to year, provided an opportunity to adjust to changes in circumstances and assumptions. Annual Plans being close to events might seem to require a smaller margin for error and to correspond more to reality. Often this was the case but the first few years of independent Bangladesh were highly disturbed and rapid and significant changes took place which added to the difficulties of economic management. Frequently plans and policies had to be modified in response to pressure of events. In this section an attempt is made to analyse the deviation of plans from reality, the response of the plans and programmes to shortfalls in resources and the effects they had on the pattern of development efforts.

To what extent did the projections of the Five-Year Plan regarding the availability of resources as well as the size and phasing of the Annual Plans materialise? What was the nature of and reasons behind the divergence between projections and realisations, or between plans and achievements? How did the size and priorities in the Annual Plans respond to changes in the availability of both domestic and foreign exchange resources? Were there deliberate over-estimations or optimistic forecasts? Were the sectoral allocations of development expenditures in the Annual Plans different from those of the Five-Year Plan? Were the provisions in the Annual Plan departed from in the course of its implementation?

The Five-Year Plan as it was implemented by successive Annual

Plans was intended to achieve two broad objectives. The first was to bring the economy back to its position of 1969/70 before the disruption caused by the war and its aftermath led to a fall in output, in the utilisation of capacity and in the dislocation of production organisation. The other objective, once the effects of the war were overcome, was to press on with the changes in the structure of the economy and to achieve levels of output and income above and beyond the benchmark (mainly pre-war) level. The overall and sectoral growth rates which the Five-Year Plan intended to achieve were relatively modest in relation to the benchmark period; but in relation to the actual levels prevailing in 1972/73 however, the Plan targets appeared very ambitious as seen below.[1]

GROSS DOMESTIC PRODUCT AND ITS COMPONENTS AT 1972/73 PRICES
(million Takas)

	Benchmark A	Actual 1972/73 B	Projected 1977/78 C	Rate of Growth: Projected over Benchmark (Pre-war)	Rate of Growth: Projected Over Actual (1972/73)
Agriculture	28,830	27,220	36,020	4.6	8.4
Manufacturing	5,200	3,298	7,310	7.1	15.4
Construction	1,840	1,444	3,260	12.1	13.7
Others	14,160	11,338	18,830		
	50,030	45,300	65,420	5.5	8.8

Income and output in 1972/73 were 10% lower than the benchmark output. About 66% of the shortfall in GDP in 1972/73 below the benchmark level was due to the fall in agricultural output and another 25% to the decline in industrial output. The attainment of the targets laid down in the Plan was critically dependent on the expectation that the major part of the recovery of the economy from wartime destruction and post-war dislocation would be accomplished by 1973/74 and with it the restoration of the levels of production and efficiency of 1969/70. All this assumed that the rehabilitation and reconstruction of the physical infrastructure, especially the repair of roads, bridges, railways and port facilities, etc., would be completed according to schedule, and that the organisational and institutional changes would be accomplished in time. It was expected that the principal task of the Annual Plan was to spell out the policy measures and institutional changes required to expedite the process of recovery of the economy.

Before the inauguration of the Five-Year Plan, a half-yearly plan for the period between January and June 1972, and an Annual Plan for the year 1972/73 had been prepared. Both these plans included large relief, rehabilitation and reconstruction components. About 70% of the expenditure scheduled for the period, January–June 1972, and 35% of that for 1972/73 were for rehabilitation and reconstruction. During the early part of 1972 more than 80% of the expenditures on reconstruction, relief and rehabilitation were for direct relief and rehabilitation of displaced persons, their transportation, feeding and temporary accommodation in relief camps. The proportion of this kind of expenditure declined to less than 40% during the following year when the reconstruction of physical assets damaged by the war became predominant. By 1973/74, the reconstruction was down to 25% of the total Annual Plan expenditures.[2] In fact, it was expected that there would be no need for any significant amount of reconstruction expenditures beyond 1973/74.

In fact, however, recovery proceeded more slowly than expected. This reflected scarcity of resources, both domestic and foreign, as well as lagging improvements in managerial and organisational efficiency. Moreover, natural calamities and adverse external economic circumstances explained elsewhere delayed recovery and growth. The gross domestic product measured in 1972/73 prices did not recover to the pre-war level until 1975/76 and that to a large extent was due to a bumper crop and good weather conditions. Agricultural income remained depressed below the pre-war level until 1975/76. Value added in the non-agricultural sector did not reach the level of 1969/70 until 1974/75. The only sectors which made consistent improvement, in terms of increase in value added, were the service sectors, i.e. public administration, trade, transport and housing services.

The provisions in the Annual Plan for development expenditures fell far short of those projected in the Five-Year Plan. During the first four years the combined total of projected development expenditures provided for in the successive Annual Plans was about Tk.27 billion as against about Tk.29 billion at 1972/73 prices provided in the Five-Year Plan. The shortfall in the Annual Plan projections, compared to the phasing of the Five-Year Plan, was larger in the earlier years than in later years. The provisions in the Annual Plan, of course, were much worse in real terms because prices had risen steeply. In real terms, the Annual Plan provisions in 1973/74 were 48% less than that projected in the Five-Year Plan. In 1974/75 it was 68% less and in 1975/76, 35%.[3]

There were two principal reasons for the difference between

provisions in the Annual Plan and the projections of the Five-Year Plan. One reasons was that when the Five-Year Plan was prepared and approved, the political authority or Cabinet did not thoroughly explore or discuss the implications of the Five-Year Plan in terms of the effort needed to mobilise resources. It was only when the Annual Plan was under consideration that the detailed policy implications of the alternative sizes of development expenditures were clearly grasped and the hard policy decisions, some of which were politically unpopular, confronted the Cabinet. Many of the policies advocated were not regarded as politically feasible, at least to the extent to which the fulfilment of the targets of the Five-Year Plan would have required. The other reason was that the principal assumption of the Five-Year Plan about the rate of recovery of production, particularly food output, and the movement of prices, both national and international, went wrong; one aspect of this was a fall in the international terms of trade. These factors are explained in later sections.

An analysis of the factors which determined the size of the Annual Plan in 1973/74 illustrates the range of considerations and policy issues which governed the size of the successive Annual Plans.[4] A consideration which naturally came to mind when deciding on the size of an annual plan was the amount of development expenditure in the previous years. The Annual Plan of 1972/73 had projected a development programme of Tk.5,010 million but realised in fact a programme of about Tk.3,978 million—a shortfall of about 20%. What lessons did the experience of 1972/73 have for the next Annual Plan of 1973/74? The pessimists argued that this was an indication of the limits of implementation capacity and resource availability in post-war Bangladesh and only a slow and moderate rate of acceleration could realistically be expected above this level. The optimists, on the other hand, argued that the experience of past years emphasised all the more the need for concerted and vigorous efforts to mobilise resources and to improve the efficiency of the economy. There was no room for slackness and the potential of the economy was greater than it seemed and was not fully harnessed in 1972/73. Moreover, the first year after the dislocation of the war should not be taken as a guide for future possibilities. Another argumment was that the targets of a plan should be more than a mere projection of past trends. A plan should aim at an improvement over past performance. Was it more appropriate to choose deliberately modest targets and to fulfil them rather than to have a larger plan which could stimulate hope and enthusiasm and provide impetus to changes in policy and institutions but which might not achieve all the targets? It was held by some that what was really important more than the fulfilment of the quantitative targets of a plan was the impact of the plan on the policies and institutions and changes in the economic structure. The Annual Plan for 1973/74, the first year of the Five-Year Plan, had to be seen in relation to the Five-Year Plan as a whole and its size and phasing. The smaller the size of the Annual Plan in the first year, the higher the rate of acceleration which would be required in the following years, if the projected total expenditures for the entire Five-Year Plan period were to be preserved. This would result in the postponement of critical policy decisions until later years; moreover, such a rate of acceleration might not be possible and sufficient momentum might not be attained if efforts in the first years were relaxed. Moreover, it was necessary to ensure that acceleration took place in real terms and not in monetary terms only; therefore, in view of the rising prices, the rate of acceleration in annual development expenditures from year to year should take into account the effect of rising prices.

In order to maintain the momentum of development efforts, the size of the annual development programme had also to be seen in relation to the size of annual development expenditures during the late 1960s, i.e. the last years before independence. It was not politically defensible for independent Bangladesh to produce Annual Plans of a size which were smaller than the average reached in the last days of the Pakistan regime. Towards the end of the 1960s annual development expenditures in Bangladesh (the former East Pakistan) amounted to about Tk.3,500 million which was roughly equivalent to Tk.5,000 million in 1972/73 prices; in the prices of 1973/74, this would have amounted to Tk.6,000 million, and this was the minimum figure which was under consideration at the time of preparation of the Annual Plan of 1973/74. The need to maintain expenditures on all ongoing projects as well as the desire not to refuse offers of new aid or to postpone expenditures on projects for which foreign assistance was already committed, provided a two-fold constraint on the minimum size of an Annual Plan. Moreover, the utilisation of foreign aided projects required counterpart local expenditures. No less than 40% of the cost of capital equipment, financed by foreign aid, was usually involved in the mere transportation of the equipment to the site of the project and its installation. Moreover, it was impossible to discuss the size of the development programme without considering the magnitude of non-development expenditures. For example, it was argued that there should be a reasonable balance between public sector development and non-development expenditures in the sense that a balance between consumption and investment expenditures was necessary for the economy as a whole.[5] During 1973/74 development expenditures were about double the amount of non-development expenditures. A lower level of development expenditures at present would have the effect of reducing the levels of consumption in future years.

Thus the lower limit to the Plan size was set by such considerations as accommodating ongoing and foreign aided projects, maintaining a growth momentum over past expenditures in real terms, providing a focal point for mobilising political enthusiasm, generating a commitment to economic development and providing a spur to increased domestic efforts for growth. The upper limit was set by the scarcity of resources or political constraints on policy measures and implementation capacity, which, it was hoped, would be elastic and would respond to a challenge. There were those who argued that it was not irrational to opt for a smaller development plan, if faulty project selection or inefficient implementation yielded little or inadequate returns on investment which were not high enough to offset the sacrifice of present consumption in view of prevailing poverty. The size of the Annual Plan of 1973/74, which was eventually decided upon, was a compromise between these conflicting considerations.

The debate on the size of the Annual Plan was mainly related to the public sector programme. There was much greater uncertainty about the performance of private investment, due partly to very inadequate data and partly to uncertainty about the success or feasibility of policy instruments designed to stimulate or regulate private investment. The main policy instrument was the provision of foreign exchange resources through the government–controlled financing institutions.

The most important bottleneck in the way of implementation of plans was the shortage of resources. During both 1972/73 and 1973/74, there were shortfalls in revenue as well as unexpected increases in current expenditures, resulting in deficits on revenue account instead of the projected surplus. It was not until 1974/75 that actual revenue receipts met and indeed exceeded the estimates and there was a surplus on current account in spite of a rise in current expenditures. In both 1972/73 and 1973/74 deficit financing was largely used to meet the deficit on current account and only to a very small extent to finance development expenditure, which was largely financed by foreign aid. It was only in 1974/75 that deficit financing was used only for the purposes of financing development expenditures. In that year also the annual target of development expenditures was realised, helped by an increase in revenue receipts and greater foreign assistance than predicted. The divergence between projections and actual realisation was not limited to domestic resource mobilisation; in 1972/73 and 1973/74 the flow of external assistance also fell below projected levels. The shortfall in the flow of foreign economic assistance was due both to a shortfall in commitment and to slow disbursements. Project assistance was disbursed more slowly than commodity assistance; in each category grants were more quickly disbursed than loans or credits. Most of the development projects in the Annual Plans were financed by loans and increasingly so in the later years. During 1972/73 and 1973/74, project assistance was about 15–16% of the combined total of new commitment and aid in the pipeline; it fell to 9–10% in the next two years, with an increase in pipelines and new commitments which could not be matched by an increase in the rate of utilisation. The latter was a function of the administrative and implementation capacity in Bangladesh, on the one hand, and rules and procedures set by the donor agencies for the disbursement of aid, on the other. The fact that the rate of utilisation of commodity assistance was higher did not help to undertake a larger development programme, except in terms of generation of local resources and helping the utilisation of existing capacity.[6]

The shortfall in resources accounted for a difference between the size of annual development expenditures, as projected in the Five-Year Plan, and its actual implementation in Annual Plans. Moreover, as sectoral allocations were worked out in the Annual Plans, divergences emerged between them and the projections of the Five-Year Plan. The reasons for this divergence were several. In the first place, it was relatively simple to provide a desirable pattern of allocations in the Five-Year Plan and quite another thing to find suitable projects and programmes for the different sectors to match the original allocations. The Annual Plan had to identify and provide for specific projects and programmes. The capacity of the different ministries to prepare and implement projects differed widely. In the second place, the pattern of commitment of aid, including pipelines inherited from the past, did not always match sectoral priorities or the time phasing of sectoral investment priorities. Some sectors were more fortunate than others. For example, the social sectors (education, health and social welfare, etc.) were weakest in terms of well-prepared projects as well as in implementation capacity. For these sectors there were delays in the preparation of projects as well as in the negotiation of foreign assistance. Moreover, changes in relative prices over time had led to variations in the relative costs of projects in different sectors, but not in the same proportion with the result that with the same physical composition of projects, financial allocations were different. Faced with a shortage of resources for investment, attempts were made to protect high priority projects of agriculture and water development as well as to continue as many ongoing projects as possible in different sectors, even if on a reduced scale. The net result of the various factors mentioned above was an increase in relative expenditures on agriculture, water, flood control, transport and communications and a relative decline on industrial and social sectors, including physical planning and housing.

A comparison between the development expenditures during the first four years of the Plan period and the Five-Year Plan projections reveals that the social sectors such as health, education, etc., as well as physical planning, housing and industry fared worse while agriculture, water, rural institutions, physical infrastructure, transport, communications and power fared better than envisaged in the Five-Year Plan.

The acute scarcity of resources and the predominance of ongoing projects restricted the possibility of wide divergence between targets of the Annual Plan and their implementation. In fact, in all the years, during the mid-year revisions of the Annual Plans, the financial resources were found to be smaller than originally estimated and had to be reallocated between the agencies and ministries, primarily on the basis of such considerations as past progress in the implementation of projects and the expected rate of performance in the future. The percentage distribution of actual development expenditures during the years between 1973 and 1976, compared to that projected in the Five-Year Plan, is shown below:[7]

FIVE-YEAR PLAN TARGETS AND SECTORAL DISTRIBUTION OF ACTUAL DEVELOPMENT EXPENDITURES
(*percentage distribution*)

	Actual Expenditures	*Five-Year Plan Targets*
1. Agricultural, rural development and water and flood control	31.6	26.3
2. Industry	15.2	18.7
3. Power, national resources, transport and communications	33.5	29.5
4. Physical planning and housing	6.3	8.0
5. Social sectors and others	13.3	17.5

Throughout the years development expenditures on agriculture, rural development, water resources and physical infrastructure constituted the largest proportion of allocations. Each year when cuts in allocations became inevitable, these sectors were spared as much as possible. Agriculture and related sectors were protected because of their importance in the economy; physical infrastructure was favourably placed in view of the relatively large proportion of ongoing projects, many of which had foreign aid committed to them, awaiting disbursement. The social sectors received the largest curtailment as resources fell short and so also the industry sector; in the latter case, aid negotiation also took a long time since there was very little in the pipeline.

To what extent did the ministries undertake development expenditures which were not provided in the Five-Year Plan or Annual Plans? Pressures were exerted by the various ministries for the inclusion of new projects and programmes at the time of preparation of the Annual Plan, even when they were not fully worked out at the time. This was considered necessary in order to accommodate the results of new analysis or changes in circumstances. During the course of the implementation of Annual Plans, the most important pressure, however, was to undertake expenditures which did not meet the various criteria and conditions which were laid down in the Five-Year Plan for the techno-economic evaluation and appraisal of projects and programmes. One such important deviation from the provisions in the Plan took the form of expenditures on unapproved projects. These projects were included in the Five-Year Plan and in the Annual Plans, but expenditures on them were not to be incurred until after their cost–benefit analysis had been finalised. They had to go through the process of approval by the inter-ministerial project evaluation committee, and eventually by the Cabinet. There were ministries which were unable to complete in time the analysis necessary for the evaluation and approval of some of these projects; yet they pressed to incur expenditures on them anyway. Frequently there were pressures to spend on unapproved projects out of funds which had remained unutilised in respect of approved projects because the latter were held up by delays and bottlenecks in implementation. In view of the low aggregate volume of development expenditures caused by shortage of resources, it was politically infeasible to allow some ministries to accumulate unspent funds at the end of each year, while at the same time to refuse funds to the ministries who, being more efficient in implementation, were able to spend but were unable to complete in time the required cost–benefit analysis. Every year there were a few unapproved projects awaiting evaluation and appraisal, on which expenditures took place under the pressure of the sponsoring ministries. However, these projects were not entirely without some kind of preliminary scrutiny and analysis, although the exhaustive analysis of the kind expected of well-prepared projects was lacking. In a few cases formal approval was given even on the basis of inadequate analysis. But in no case were important and large projects involved.

Another instance of deviation from the Plan was when the rate of implementation of a project was speeded up beyond the original level postulated in the Plan. The most outstanding case was the programme for the procurement of low-lift pumps and tubewells to be installed for irrigation. There were a number of recommendations in the Five-Year Plan designed to ensure a more economical and efficient utilisation of low-lift pumps as well as of deep and shallow tubewells. The extension of irrigation facilities was conditional upon the Ministry of Agriculture implementing these recommendations; however, the ministry went ahead installing pumps and tubewells without fulfilling these conditions. The result was an increase in irrigation pumps and wells without a corresponding increase in the irrigated acreage. Moreover, the ministry speeded up orders for imports of low-lift pumps in excess of the rate laid down in the Annual Plans, although not in excess of the total numbers provided for the entire Five-Year Plan period. They made commitments involving suppliers' credit and later on pressed for its substitution by financing through foreign assistance, specifically project assistance from the various bilateral sources from which contracts for purchase were made. For reasons explained elsewhere, there was great pressure of demand especially on the part of the more prosperous farmers as well as great competition between regions and localities for irrigation pumps beyond what was justified on social cost–benefit considerations. The Ministry of Agriculture was greatly enthusiastic in responding to this demand, partly because it considered the extension of irrigation to be the quickest method for increasing food production. Moreover, there was a powerful lobby for the sale of irrigation equipment amongst the traders or importers as well as amongst the local agents of overseas manufacturers.

Yet another instance of pressure for deviations from Plan allocations concerned the construction of local roads. There were a large number of local roads, some of which had been started in the pre-independence period but only partially completed; there was a clamour for more funds to be allocated to them and to meet the increasing costs of road construction. The roads were often technically defective and badly designed; they seldom outlived more than one season of floods. The Five-Year Plan recommended that the non-arterial or "feeder" roads should be constructed by the local government institutions with land and labour to be mobilised locally and paid for either in kind or cash; the central government was to provide material inputs (cement and other construction materials) as well as technical and engineering help for design and construction. This particular institutional arrangement was intended to encourage the mobilisation of local resources for projects which brought immediate benefits to the locality. It was also expected to ensure that responsibility for mobilising local resources would generate pressure for a more careful analysis of economic benefits to be derived from them. The Ministry of Roads, however, pressed for an allocation of funds for meeting the total costs of roads and not merely for the central government's contribution; the ministry also undertook the construction of local roads through centrally appointed contractors rather than through the local government institutions as advocated in the Plan.

Faced with the demand for the construction of a very large number of local roads, on the one hand, and scarcity of funds, on the other, it was difficult for Annual Plans to select among the competing claims for road construction all over the country. It was decided instead that a certain block allocation of funds would be made to the Ministry of Roads, which would select the indvidual roads to be constructed in particular areas. The allocation of funds by the ministry between competing roads was basically determined by the relative bargaining power of the different regions and localities. The result was a wide dispersion of the limited resources among a large number of roads, each of which could only be partially constructed and was technically deficient. The engineering and supervisory capacity of the central ministry was quite inadequate to directly supervise and construct such a large number of local roads all over the county.

To cite another example of the faulty implementation of projects, an important project in the Five-Year Plan was the multi-purpose integrated rural health centres for the provision of both health and family planning services. Since this was a new experiment, initially only a few centres were to be fully established; a careful evaluation of their performance was to determine the pace and pattern of their further expansion, if necessary with modifications and alterations in the light of experience. This scarcity of resources also dictated that only a limited number of centres should be constructed in the first phase. In fact, instead of being concentrated in a few selected centres, resources allocated for the health centres were spread over the entire country by the ministry of health in a vain attempt to provide as many prospective health centres as possible. Insufficiency of resources resulted in each health centre receiving allocations which were much less than was necessary to complete the construction or to develop the full range of their activities.

Previously the health and family planning services were separate and independent activities. The Plan provided for a new approach through their integration in stages; however, it left room for revision and modification of the programme in the light of the evaluation of the lessons of experience. The Ministry of Health and Family Planning was inflexibly committed to this approach. The family planning services disliked the integration, partly because they would lose their independence in the integrated organisation; the latter would not only be placed under the Ministry of Health but also was most likely to be dominated and led by the health and medical services. The Health Ministry was very unhappy with the lack of unequivocal commitment on the part of the government towards the new approach. By committing funds, however small, to all the health centres at the same time all over the country, it felt that it could make the government's commitment to the programme irrevocable and therefore could forestall the possibility of its possible revision. But the financial allocations were inadequate and, therefore, the construction programme was thinly spread out all over the country. This frustrated not only the original purpose of "learning by doing" but also prevented even a limited number of health centres from being completed and fully staffed.

The Five-Year Plan and the Annual Plans recommended in considerable detail the specific policies and institutions for the implementation of the sectoral programmes. In many cases the organisation and structure of the agencies or institutions required for the implementation of projects, including their methods of operation, were specified. This was necessary partly because many of the agencies and institutions were new and partly because of the change in the organisational structure and operational procedures, which was suggested in the Plan. For example, it was suggested that a large part of investment in housing, including housing for government officers, was to be undertaken by the housing co-operatives, which were to receive technical and financial assistance from the government. However, the organisation of co-operative societies for house-building was a time-consuming process; they required considerable initiative and enterprise on the part of the prospective users of houses as well as the ministries of co-operatives and housing. The Housing Ministry and its agencies traditionally used to undertake direct construction for government officers. Faced with a very great pressure for providing housing, it chose the soft option; it undertook the construction of houses on its own and abandoned the idea of concentrating on housing co-operatives.

In the field of education, the Plan recommended limited growth in the enrolment in the university, especially such branches as liberal arts and humanities. Similarly, the expansion of liberal arts colleges was to be restricted. In fact, the expansion of university education provided in the Plan in these branches was already considered excessive; compromises had to be made due to political pressures and political commitments made in the field of higher education. It was, therefore, expected that since the Plan itself had made concessions to political pressures, further departures from the Plan priorities would not be necessary. However, in the course of the implementation of the Plan, the rate of expansion of university education was accelerated far above the target set in the Plan. The target which was set for the expansion of enrolment provided in the entire five-year period was reached in the first year. This was a major deviation from the Plan. Moreover, the government's decision to provide resources from the government budgets for the entire needs of the privately financed colleges, many of which were liberal arts or general science colleges and were previously financed only partly by the government, made a significant step towards the expansion of the public expenditures on higher education in a way which was contrary to the Plan projections. Also, several new experiments for economising on the cost of education such as multiple shift working of schools, etc., as suggested in the Plan were not undertaken.

A large component of ongoing projects in the annual development plans and an acute shortage of resources made significant departures in actual implementation from provisions of the Annual Plan almost impossible. However, as discussed above, the Plans had laid great emphasis on recommendations for changes in organisation, institutions and policies in almost all sectors with a view to improving their performance to formulate and implement policies and projects and to increase their efficiency of operation; it was with respect to these recommendations that departures from the Annual Plans were more frequent and glaring than with respect to any distortions in the sectoral priorities of investment plans.

(b) Annual Plans and Food Production

The implementation of the Annual Plan was highly dependent on the performance of agriculture, on developments in foreign trade and on movements in prices which were often violent. Agriculture constituted 60% of GNP and gave employment to 70% of the labour force; it was of overwhelming importance. Rice production occupied by far the most dominant position. Eighty per cent of the cropped area was devoted to rice; jute, the next most important crop, covered only 7–8% of the cropped area. The output of both these crops was subject to fluctuations owing to variations in weather conditions. Fluctuations in the production of foodgrains had important consequences for income, employment and the balance of payments. The variations in the relative prices of jute and rice had important implications for their relative output and exports. This section discusses how the Annual Plan attempted to provide for fluctuations in the output of food and how it responded to the consequences of such fluctuations on the rest of the economy.

Food production suffered a serious setback during the year 1971/72 due to the dislocations of war, aggravated by drought which adversely affected the monsoon (Aus) crop harvested in the middle of the calendar year, 1972. The output of rice was about 10 million tons as compared with 12 million tons in 1969/70, the normal year before independence. The target of rice production in 1972/73, based on the availability of fertilizers, pesticides and high-yielding seeds and on the assumption of normal weather, was optimistic. Output was expected to reach the level attained in 1969/70.

The achievement of the Annual Plan fell far short of expectations. The year 1972/73 witnessed the most serious drought in recent history which affected both the Aman and Aus crops very severely. While the drought was the main cause of shortfall, slow implementation of the development programme, especially of the procurement and distribution of seeds and fertilizers, also contributed to the decline. The winter crop, Boro, also suffered due to early rains at harvesting time in 1973. Actual output of rice during 1972/73 was 25% below the projected level. Imports during 1972/73 amounted about to 2.7 million tons, far in excess of what would have been the case if the programming of imports was based merely on the statistical gap between the expected or projected output and the estimated consumption requirements. About 2.6 millions tons of foodgrains were distributed through the public distribution system. The per capita consumption of foodgrains for the year was about 358 lb, much less than 388 lb consumed in 1969/70.[8]

For the year 1973/74, production of foodgrains was first estimated at 12.6 million tons as against 13 million tons for 1972/73 and total import requirements were estimated at about 1.5 million tons, including 300,000 tons for an increase in government stocks. By the end of the calendar year 1973, forecasts for output were revised downwards to 11.8 million tons and the import requirements were raised to 2.20 million tons, including public stocks of 400,000 tons. While the actual food production was close to the revised projection, the actual imports fell short by ½ million tons. Consumption could only be increased only slightly from the very low level of 1972/73 at the expense of a decline in stocks.

The year 1974/75 started with the most optimistic forecast that output of foodgrains would exceed 13 million tons; it could not be foreseen that the country would be hit by one of the worst floods in thirty years. There was extensive damage to crops, aggravated by a failure to provide inputs needed for increased domestic production. In the post-floods period, forecasts of output were revised downwards and imports revised upwards. While actual imports matched estimated requirements, they were not available for distribution through the rationing system during the year, because imports did not arrive in time to meet the monthly requirements of the public distribution system. It was not possible to compensate for the effects of the time lag in the arrival of imports by drawing down on stocks. At the beginning of the year 1974/75, the foodgrain stocks were down to 230,000 tons compared with averages of 500,000–600,000 tons before independence when the population and the requirements of the public distribution system as well as the food gap were smaller. In practice when stocks were reduced to 200,000 tons or so, no more than half of this was actually available for distribution because a large part of stocks, as estimated statistically, include foodgrains on ships, in transit in the country and in the central storage depots. Unless foodgrains were available in local depots, they could not be passed on to the local ration shops and stocks were not in practice available for distribution to the public. The most critical months of shortage were the months of September and October and the domestic output which would have been available for these months was severely damaged by floods. In 1972/73 and 1973/74 imports during these two months amounted to 435,000 tons and 550,000 tons respectively. During 1974/75, imports during these two months together amounted to about 100,000 tons even though in the subsequent months, the flow of imports was accelerated. The uneven or irregular arrival of imports, coupled with the lack of domestic stocks, resulted in famine and death from starvation during the last quarter of 1974. Owing to the late arrival of imports, domestic consumption was considerably less during this year and stocks were built up at the end of the year 1974/75, at about 0.68 million tons which were the largest stocks ever held in the post-independence period. The per capita consumption of foodgrains in 1974/75 was about 339 lb, which was the lowest in recent memory,[9]

The year 1975/76 not only began with a large stock of foodgrains but ironically faced very favourable weather conditions. The projected output of food was about 13 million tons; imports were estimated originally at 2 million tons to provide a large addition to stocks and to allow an increase in per capita consumption after years of deprivation.

As has been pointed out earlier, apart from adverse weather conditions, the shortfall in the output of foodgrains below the target level was partly due to the slow rate of implementation of the agricultural development programme. This in turn was due to the shortage of domestic and foreign exchange resources. Foreign exchange resources were seldom available in time in order to be able to take advantage of the opportunity of favourable purchases in world trade or to undertake advance purchases of fertilizers and pesticides, etc., which could have ensured an even supply of inputs to the agricultural sector. To this was added the inefficiency of the administrative and institutional framework which aggravated the effects of resource shortage. The shortfall in the distribution of fertilizers, the most critical input for increasing rice output, is shown below.[10]

DISTRIBUTION OF FERTILIZER
(000 tons)

Year	*Target*	*Achievement*
1972/73	425,000	375,000
1973/74	473,000	373,000
1974/75	500,000	340,000
1975/76	600,000	—

The import and distribution of fertilizers was undertaken by the public sector up to the thana level in the rural areas. The thana was the third stage in the administrative hierarchy of the country from below, the first stage being the village. Between the government stores at the thana level and farmers in the villages who were the actual users of fertilizers, several agencies such as farmers' co-operatives, local government agencies and private traders intervened with varying degrees of participation and control, which to make matters worse, changed over time. There was a lack of co-ordination between the various agencies, public and private, involved in the distribution of fertilizers with resultant delays in the distribution of fertilizers to the farmers.

An illustration of the nature of the dislocation which did occur from time to time in the system of distribution of fertilizer is provided by the experience of 1974/75.[11] There was firstly the breakdown of the Ammonia control room in Ghorasal Fertilizer, the biggest fertilizer factory in the country, in September 1974, just the period when fertilizer was critically needed for the rice crop in the subsequent months. Added to this was the uncertainty in the arrival of fertilizer procured from external sources. The floods in late 1974 caused dislocation in the system of internal transportation and distribution of fertilizer. As if this was not enough, the sale of fertilizers was suspended by the government for stock-taking just after the explosion in the Ghorasal factory, although this was the peak season for application of fertilizers to the Aman crop. The embargo on its sale was not lifted until the first week of October, even though it was decided to sell fertilizers only to those farmers who used high-yielding varieties of seeds.

There was a shortfall in the implementation of the irrigation programme, which was the next most important component of the programme for accelerated rice production. There was a delay in the sinking of tubewells and low-lift pumps as well as in multi-purpose irrigation, drainage and flood control projects. In the latter category, the targets of additional irrigated acreage were respectively 60,000 acres in 1973/74 and 40,000 acres in 1974/75. The actual areas irrigated in each of these two years were no more than 30,000 acres. Under the first category of irrigation programme, i.e. the low-lift and tubewell programme, the cumulative irrigated acreage was expected to be 1,380,000 acres by 1973/74 and 1,322,000 acres by 1974/75 whereas the actual irrigated acreage was no more than 1,136,000 acres and 1,147,000 acres in 1973/74 and 1974/75.[12]

The fluctuations in weather conditions and lags in the implementation of development programmes affected, in addition to foodgrains, an important agricultural crop, which had important implications for the Annual Plan. This was the fluctuation in the output of jute—the principal export commodity. The final estimate of jute output was available at the beginning of the financial year since it was a monsoon crop, competing with the Aus crop. The development programme for raw jute, popularly known as the intensive jute cultivation scheme, using improved seeds, fertilizers and modern cultural practices, was not as yet under way. Therefore, the unusual variations in the output of raw jute were dependent on changes in weather conditions and on the relative prices of rice and jute, which determined the allocation of land between the two competing crops.[13]

PRODUCTION AND ACREAGE OF RICE AND JUTE

	Jute output (million bales)	*Acreage (million acres)*	*Rice Output (million tons)*	*Acreage (million acres)*
1972/73	6.5	2.2	10.0	23.8
1973/74	6.0	2.0	11.8	24.4
1974/75	3.9	1.4	11.2	24.2
1975/76	4.3	1.3	13.2	25.6

The impact of fluctuations in the outputs of both jute and rice had implications for the balance of payments since one was a foreign exchange earner and the other was a foreign exchange saver. The percentage fluctuations in the output of foodgrains had a magnified impact on the import requirements and hence on the balance of payments because fluctuations in food occurred in the context of a rising population, which increased the consumption requirements of food from year to year at least at the same rate as the rate of growth of population.[14] Apart from the direct impact on the balance of payments, fluctuations in food output affected the cost structure in a pervasive way because food prices were an important contributing factor to general price movement. Apart from these general effects scarcity and fluctuations in the availability of food severely restricted the labour-intensive development programmes such as the rural public works programme.

The above narrative brings out a few important features of the food problem in Bangladesh. Between 1972/73 and 1973/74 total rice output fluctuated by 18%, whereas between 1973/74 and 1974/75, it fluctuated by 5%. The Aman crop, the major crop, which was mainly dependent on rainfall, rose by 20% between 1972/73 and 1973/74 and fell by more than 10% between 1973/74 and 1974/75. Similarly, the Aus crop, which was predominantly a monsoon crop, fluctuated by 20% between the earlier years and by 3% between the latter years. These wide fluctuations were due to variations in weather conditions, specifically due to drought in 1971 and 1972 and floods in 1974. But then fluctuations in weather conditions mainly affected the Aman and Aus crops, two rice crops which were predominantly dependent on rainfall rather than irrigation. The output of the winter crop, i.e. Boro, which was dependent on irrigation rather than rainfall, was more stable but then it constituted no more than 20–22% of the total annual output of rice. An extension of the irrigated rice acreage would, therefore, have contributed to a greater stability of production. While the extension of irrigation would have reduced the uncertainty and fluctuations in output, originating from variations in weather conditions, it would increase the dependence on organisational and institutional factors. Any failure in these directions would have increased the influence of man-made, as against natural, factors in determining the domestic output of food. The interruptions or irregularities in the supply of fertilizer or in the supply of irrigation water owing to technical, financial or organisational factors had also in the past caused fluctuations in the output of rice. To the extent that the supply of the agricultural inputs was dependent on a regular flow of imports, this in turn was affected by the fluctuations in foreign exchange earnings and in the flow of economic assistance.

There was considerable seasonal variation in the supply of foodgrains in the domestic market. Immediately after harvests the market supply was large and prices were low. Since there were three rice crops in a year, there were three harvesting periods during which food supplies in the market were relatively plentiful. But then in view of the fact that these crops were of different sizes, the magnitude of their impact on the market supply at the time of harvesting differed widely. The leanest months or months of greatest shortage were May/June and September/October. The largest crop, i.e. Aman, was harvested in November/December and flow of this output into the market and into the consumption stream continued at a diminishing rate until the middle of the next calendar year. It was augmented by the output of Boro (winter crop) harvested around February/March. May/June were also lean months when there was a relative shortage of market supply before the next Aus crop was harvested in July/August and came on to the market. By September/October the stocks of all the previous crops were generally exhausted. In view of the seasonal variations in the supply of foodgrains in the market, it was necessary to programme the supply of imports in such a manner as to meet the consumption requirements of foodgrains in different season. The public distribution system was so designed as to supply a larger quantity of foodgrains in the "lean" months than in the rest of the year.

Seasonal variations in the domestic supply of foodgrains could be offset by holding domestic stocks of foodgrains and by releasing them in the open market or under the rationing system according to the varying requirements of consumption. Since there was an overall shortage, the domestic stocks of foodgrains would only have been built up through imports at the beginning of the year. If this was possible, it was not necessary to programme variable quantities of imports in the different seasons of a year; variations in the domestic stocks for foodgrains would be an adequate substitute. Lack of resources prevented Bangladesh from building up an adequate reserve of stocks to serve as a cushion against the variations in domestic supply as between the differnt seasons or the different years. Neither did she have enough foreign exchange resources to import food in time for meeting the seasonal variations in domestic supply or unexpected shortfalls in supply. In all the years under review, actual production of foodgrains in Bangladesh fell short of the projections. Similarly, actual imports fell short of projected imports. The latter was due to both shortages of foreign exchange and a rise in the world prices of foodgrains starting in 1973.

The year 1974/75 provided a tragic example of the fact that even when the overall annual supply of foodgrains, from imports and domestic production, was adequate for consumption needs, the failure of imports to arrive in time to meet the requirement of consumption in lean months, in view of small publicly-held stocks, resulted in famine and death from starvation. The floods in 1974 affected partly the Aus crop but mainly the Aman crop which was harvested in November/December in 1974 whereas the incidence of famine was most severe in September/October 1974. The overall net domestic availability of foodgrains, from domestic production and imports for the year as a whole, was about the same in 1974/75 as in 1973/74. It was the late arrival of imports to meet domestic shortages in time which accounted for the fact that even though the total amount of foodgrains available for public distribution was 2.6 million tons, actual distribution was only 1.76 million tons and the closing stocks at the end of 1974/75 were the highest in the post-independence period, i.e. 0.68 million tons.

In view of the predominance of the output of rice in the agricultural sector, fluctuations in rice output determined the fluctuations in the agricultural sector which in turn had a dominating influence on the growth and fluctuations in the economy as a whole as is seen below.[15]

ANNUAL PERCENTAGE CHANGE IN OVERALL AND SECTORAL GNP

	1972/73 / *1969/70*	*1973/74* / *1972/73*	*1974/75* / *1973/74*	*1975/76* / *1974/75*
Total GDP	−12.0	8.7	2.0	12.0
Agriculture	−14.5	11.3	−2.0	13.8
Industry	−23.0	14.7	−1.3	5.3
Construction	−27.5	−56.8	139.0	19.0
Power and Gas	18.7	76.1	2.3	19.0
Transport	0.0	9.0	0.0	14.0
Housing Service	2.9	2.8	2.8	4.0
Public Administration	10.4	39.4	26.6	12.0
Banking and Insurance	26.9	3.2	3.3	6.0
Professional and Miscellaneous Services	3.0	1.0	1.0	4.0

The rate of change in the total GDP was highly correlated with that in the agricultural sector; the rate of change in income originating in the industrial, trade and transport sectors was highly correlated with the rate of change of output in the agricultural sector. This was because industry was based predominantly on activities processing agricultural raw materials, whereas trade and transportation activities to a large extent related to the movement of and trade in agricultural commodities. During 1974/75 it was a combination of a very considerable increase in construction activity and the public administration services which kept the rate of growth of the economy from being either zero or negative in response to a fall in agricultural income.

(c) Annual Plans and Foreign Trade

Estimates of foreign exchange resources were a key ingredient in the preparation of the Annual Plans. It proved in fact almost impossible to forecast developments in trade, even for twelve months ahead. Earnings from the exports of jute and jute products were always uncertain and it was not easy to forecast what assistance could be counted upon for the year as a whole. By themselves these difficulties would have been enough to have contended with, but they were magnified almost beyond recognition by violent swings in world trade and prices which reacted with devastating effects on Bangladesh's position. The lesson to be learnt from this experience was that no planning effort could expect to ensure smooth working of an economy when violent external shocks were experienced and there were no reserves to cushion the shock. The Annual Plans, far from being an exercise in orderly planning, became for a time devices to reduce the worse effects of unforeseen events. Inevitably this imposed enormous strains on the administrative structure and tested its ability to deal with unpredictable events to the full.

Foreign exchange shortage was a predominant constraint on the development efforts of Bangladesh. Export earnings were stagnant at a time when the terms of trade deteriorated rapidly. The volume of imports was totally inadequate to meet the minimum requirements of consumption and to provide for the raw materials needed to utilise productive capacity. It was not possible to sustain the rates of growth laid down in the Five-Year Plan or for that matter in each of the Annual Plans. The behaviour of principal commodity exports is shown below.

MERCHANDISE EXPORTS*

	Raw Jute (q) million bales	*Jute (v) $ million*	*Jute (q) 000 tons*	*Goods (v) $ million*	*Tea (q) million lb*	*Tea (v) $ million*
1972/73	2.83	126	410	179	45	9
1973/74	2.66	124	440	194	47	13
1974/75	1.55	86	340	213	52	22
1975/76	2.00	108	430	183	53	21

* v = value in million dollars; q = quantity of raw jute in million bales, that of jute goods in million tons and that of tea in million lb.

There was a steady decline in the quantity of raw jute exported; exports of jute goods fluctuated from year to year but failed to grow. The slow growth and fluctuations in exports of raw jute and jute goods reflected difficulties in domestic production, in internal marketing and transportation as well as in external shipping. Basically the slow growth of jute exports was caused by difficulties of supply until 1974 when recession hit the industrialised countries. In none of these years did the production and export of raw jute or jute goods regain the level reached in 1969/70. The highest quantity of raw jute exported in the post-independence period was 2.83 million bales as against 3.50 million bales in 1969/70. The highest quantity of jute goods exported in 1973/74 was 440 thousand tons as against 498 thousand in 1969/70. The recession and contraction of demand in the industrial countries adversely affected both the quantity and price of exports of raw jute and jute goods. At the same time the accumulation of stocks during the period 1973/75 was partly due to inadequate sales promotion by state trading agencies (exports of raw jute and jute goods were nationalised after independence) and deficiencies in grading and shipping.[16]

The shortfalls in foreign exchange earnings (including invisibles) below levels projected in the Annual Plans may be seen from the table below.[17]

FOREIGN EXCHANGE EARNINGS (INCLUDING INVISIBLES)
(*million dollars*)

	Projected	*Actual*
1972/73	325	375
1973/74	425	393
1974/75	474	397
1975/76	512	354

In every year except for the first, actual exchange earnings fell short of projected values. During 1972/73, projections of exchange earnings were deliberately biased downwards in view of the shipping and transportation bottlenecks, which were particularly significant in this year. Exports fell short of projections except during 1972/73 when they recovered from the effects of the war of liberation more quickly than expected. In fact, there was a gap between projections and actual imports in all the years between 1973/74 and 1975/76. While actual commodity exports were over 90% of those projected for 1973/74, the percentage dropped to 70% in 1975/76. Jute output had fallen and export demand was low.

Another source of instability in the balance of payments was fluctuations in the flow of foreign assistance. The importance of foreign economic aid in the implementation of the Annual Plan is seen not only from the fact that foreign economic assistance financed 75–80% of the total imports of Bangladesh. Because taxes on imports provided a large part of government revenues, fluctuations in export earnings and foreign assistance had a marked effect on the financial position of the government and hence on the magnitude of public saving available for investment. Moreover, food aid provided counterpart domestic funds for financing development expenditures. It was a way of mobilising resources from the private sector for financing expenditures in the public sector. This was also true of that portion of the commodity aid which was sold to the private sector. Thus food and commodity aid sold to the private sector served as a mechanism for augmenting the resources of the public sector. The value of foreign assistance received in each of the years 1972/73 is shown below.[18]

FOREIGN ASSISTANCE
(*million dollars*)

	Non-food Assistance	*Total (including food aid)*
1972/73	376	559
1973/74	247	481
1974/75	545	918
1975/76	557	905

While exchange earnings stagnated, foodgrain imports constituted a growing proportion of total import payments, rising from 35% to 40% during 1972/73 and 1974/75; it was only in 1975/76 that with a good harvest the proportion fell to about 27%. In all the early years imports of foodgrains exceeded the projected amounts by a wide margin. In 1972/73 payments for food imports were more than twice the amount projected at the beginning of the year. During 1973/74 and 1974/75 respectively, actual payments exceeded the amount originally projected by 50–60%. It was not until conditions eased in 1975/76 that for the first time payments for food were less than the projections.

Faced with failing foreign exchange earnings and unexpected demands for food imports the planners were placed with a difficult choice. The choice lay between securing increased foreign assistance, if possible, and curtailing non-food imports. In fact, it was necessary to resort to both. Non-food imports had already been cut to the bone. A few essential consumer goods and industrial raw materials such as textiles, cotton yarn, raw cotton, crude and refined petroleum, fertilizer and cement constituted between 30–40% of the total non-food import bill; capital goods accounted for a further 17–25% of imports and the rest was made up of the vast bulk of industrial raw materials and essential consumer goods. The composition of imports, dominated as it was by items of essential need, restricted the ability of the economy to adjust in response to changes in the availability of foreign exchange. The opportunity to squeeze imports at times for foreign exchange shortage was very limited. One consequence was that a very low priority had to be given to development goods.

During 1972/73, non-food imports fell short of the projected level by about one third, even though exchange earnings were 16% higher than projected. There was also a shortfall in external economic assistance, and in addition, there was the need to build up foreign exchange reserves, which were non-existent at the time of independence. During 1973/74 an attempt was made to protect non-food imports from cuts; but since food imports were higher than projected, there was an unexpected rise in total import payments. It was necessary to borrow short and run down foreign exchange reserves.

The year 1974/75 saw the greatest strain on the balance of payments. Jute exports fell, food imports rose to replace domestic output lost by floods, import prices rose steeply. To make matters worse, this was at a time when an increase in the real volume of imports was desperately needed to replenish stocks and keep production moving. A large deficit in the balance of payments was incurred but the volume of imports fell far short of requirements reflecting the large (74%) increase in the price of non-food imports that had taken place between 1972/73 and 1973/74.[19] The real volume of imports was considerably less than the monetary value of imports. The decline in the real value of non-food imports is shown below. The fall in the real volume of imports was about 12% in 1973/74 compared to 1972/73; real volume of imports was about

NON-FOOD IMPORTS
(*million dollars*)

	Projected	*Actual (current prices)*	*Actual (constant prices of 1972/73)*
1972/73	642	406	—
1973/74	626	609	350
1974/75	599	858	493
1975/76	1,200	928	533

20% higher in 1974/75 compared with 1972/73, and between 1974/75 and 1975/76 increase was barely 10% as seen below.[20]

The flow of foreign project assistance, especially the non-food commodity and economic assistance increased over time. Limited administrative and implementation capacity restrained any large increase in project assistance over short periods; food and commodity assistance could be more quickly committed and disbursed. Growing needs and increases in prices meant that each year higher and higher levels of foreign economic assistance had to be negotiated to cover the balance-of-payments gap as shown in the following table. In real terms the increase in foreign assistance was much less than the figures might suggest because import prices increased by 50% between 1973/74 and 1974/75 and by another 10% between 1974/75 and 1975/76. There were marked divergences between projections and actual amounts of foreign assistance; commodity assistance in 1974/75 and 1975/76 turned out to be higher than expected because additional assistance was given in the face of the balance-of-payments crisis emerging in these years. Project assistance, however, fell short of that predicted in all the years.[21]

FLOW OF ASSISTANCE: PROJECTED AND ACTUAL
(*million dollars*)

	Projected	*Actual*	*Projected*	*Actual*	*Projected*	*Actual*	*Projected*	*Actual*
Commodity Assistance	337	376	211	109	204	400	304	400
Project Assistance	—	—	150	138	159	145	167	157
	337	376	361	247	363	545	471	557

The problems created by the limited amount of imports that could be purchased can be appreciated by considering the quantity of imports of a few important items during the years between 1972/73 and 1975/76. Import requirements of edible oil, an essential item of mass consumption, were estimated to be about 100,000 tons against actual imports of only half that amount in all the years; imports of raw cotton and cotton yarn varied between 20 million and 30 million bales as against the estimated requirements of almost 46 million bales required to provide minimum quantities of coarse cloth, an essential item of consumption, and utilise the capacity of the domestic cotton textile industry; imports of cement varied between 60% and 50% of the estimated requirements. Bangladesh had a very small domestic supply of cement and was heavily dependent on imports. At times cement was so difficult to get hold of that construction came to a standstill; at all times development was held back because the supply was inadequate. The total import requirements of crude petroleum and petroleum products was estimated to be almost 1.6 million tons but actual imports varied between 1.0 million and 1.3 million tons. The shortage of kerosene was a strong cause of discontent for the common man. Most cooking was done with kerosene and scarcity meant that it was not only difficult to get but often expensive as well. The generation of electricity, the operation of transport facilities and irrigation equipment were all dependent on the supply of diesel oil. Industries using oil in large amounts such as the cement, paper and newsprint industries suffered a serious setback.

The impact of fluctuations in foreign exchange resources was aggravated by uncertainty or lack of timely availability of supplies in the world market. The prime examples were provided by variations in the supply of crude oil and rock phosphate, the raw material for the fertilizer industry. Bangladesh had only one oil refinery which was dependent on the import of a particular variety of crude oil. Its regular supply in the pre-independence days was ensured by long-term agreement with a foreign oil company, which was interrupted at the time of independence. Arrangements for the procurement of crude oil through direct negotiations either with an international company or with one of the OPEC governments took time to be finalised. In the meantime, purchases in the international market through open tenders could only be made sporadically. The shortfalls or fluctuations in the output of the oil refinery could not be met by a corresponding increase or variation in the imports of refined petroleum products. Moreover, a sequence of "stop and go" or frequent interruptions in the operation of the oil refinery, caused by irregularity in the supply of crude oil not only upset the schedule of repair and maintenance operations in the oil refinery but also resulted in the deterioration of the quality of the equipment and machinery. Thus short-run fluctuations in the supply of imports resulted in long-run damage to the oil refinery. Again, there was the example of the newly established triple super phosphate factory, which was based on a particular type of rock phosphate; the alternative variety of phosphate caused excessive corrosion and damage to equipment. It took several years before rock phosphate could be procured on a regular basis; in the meantime, the factory could not be commissioned. To make matters worse, the price of rock phosphate—world trade of which was controlled by two or three suppliers—quadrupled in line with general inflationary trends in world trade. Lack of foreign exchange reserves prevented the building up of a domestic stock by making purchases when import prices were low. Aside from the irregularity of supply of imports and world-wide scarcity of particular commodities which were badly needed by Bangladesh, the general foreign exchange shortage and overall scarcity of imports resulted in widespread under utilisation of industrial capacity. Beginning in the last quarter of 1973, Bangladesh also suffered from the imposition of export controls in the exporting countries on a few critical imports of Bangladesh such as soya bean seeds or oil and nitrogenous fertilizers, etc.

The balance-of-payments constraint and shortage of imports reached a critical stage in 1974/75. During this year the deficit in the balance of payments was 50% higher than had been expected, the combined result of floods, world-wide inflation and the fall in the terms of trade, Bangladesh was placed by the UN on the top of the list of countries most severely affected (MSA) by world-wide inflation; she received emergency assistance from the UNEO (UN emergency operation for helping the most severely affected countries), borrowed from the IMF under the second credit tranche as well as from the IMF Oil Facility which was intended to help meet the higher cost of crude oil and refined peroleum. She also borrowed short-term loans from the commercial banks abroad. Earlier in 1972/73 borrowings were made under the Compensating Financing Scheme of the IMF, to compensate for the shortfall in export earnings below the normal preindependence level. In 1973/74 the first credit tranche of the IMF was utilised. By 1974/75 all the easily acceptable sources of foreign borrowing had been exhausted. It was not possible to borrow more from commercial channels or the international money market because of limited credit worthiness. By the third quarter of 1974 Bangladesh had to launch an intensive drive to mobilise long-term foreign assistance, to speed up the commitment of foreign aid programmes, which were being negotiated and to accelerate the rate of disbursement of already committed aid. At this point of crisis Bangladesh was on the verge of defaulting on her outstanding commitments for payments for imports, for which there were no funds. Liabilities on account of letters of credit already negotiated for the purchase of imports exceeded foreign exchange reserves. Letters of credit had to be cancelled in addition to outstanding orders for imports. The country had to live from one quarter to the next in search of foreign exchange resources from abroad. Receipts and payments of foreign exchange were scrutinised every quarter and often every month to cut down imports already planned, or to postpone payments on imports already contracted for, until new assistance was available.

Even after the strictest economy in programming imports,

payments for actual imports in 1974/75 reached a figure of 1,507 million dollars in current prices (compared with the projection of 1,021 million dollars for the same year), as a consequence of the rise in import prices and a fall in the domestic supply of foodgrains. It was necessary to renegotiate letters of credit worth 203 million dollars which fell due in 1974/75; payment was postponed to the following year with the expectation that a larger flow of foreign assistance in 1975/76 would enable such payments to be met. Movements in foreign exchange reserves and short-term borrowing indicate the nature of the balance-of-payments crises which confronted Bangladesh in 1974/75, as seen below.[22]

INTERNATIONAL RESERVES AND OUTSTANDING IMF AND COMMERCIAL BORROWING
(million dollars)

		Borrowings	
	Reserves	*Foreign Banks*	*IMF*
December 1973	149		75
March 1974	97		75
June 1974	60		111
September 1974	40		124
December 1974	103	26	162

Apart from the shortage of foreign exchange resources for financing imports, other causes of an administrative nature also affected the flow of imports. There were serious delays in the issue of licences to industrial enterprises because industrial entitlements (import permits for the industrial units for the imports of raw materials) had to be revised in order to take into account changes in circumstances following independence. Actual arrival of imports was often delayed by time lag between the actual allocation of funds and the issue of licences, between the issue of licences and the opening of letters of credit, between the opening of letters of credit and the retirement of letters of credit and between the retirement of letters of credit and the arrival of imports.

In response to a request from the government an informal meeting of the creditors of Bangladesh was held in the third quarter of 1974 under the auspices of the World Bank, to explore the possibilities for emergency assistance to Bangladesh. Subsequently, a formal consortium of creditor countries, called the "Aid to Bangladesh Group", was established under the auspices of the World Bank; its first meeting was held in October 1974. Until 1974 Bangladesh was not enthusiastic about the formation of a consortium of donor countries. The resistance was partly because of her apprehension that this would enable the donors to combine to exert pressure on and influence the pattern of her economic policies and indirectly on her political choices, both internally and externally. Her experience with the flow of foreign economic assistance during the years between 1972/74 did not make her feel that a consortium could have done better in this respect. Moreover, there was also the belief that whatever co-ordination of the flows of assistance, by types and by donors, was necessary could be handled by the Bangladesh government in bilateral consultations with donors. The government wanted to retain and expand its economic and political relationship with the socialist bloc of countries; it was felt that joining a consortium which was sponsored by and composed of the Western donors could be interpreted as a move closer to the Western bloc, thus casting doubt on her neutrality. Undoubtedly a part of the doubts and uncertainties of the government was due to the fact that as a newly independent country Bangladesh was still in a stage of transition in forging economic and political relations with other countries, both in the developed and developing world; it wanted to keep is options open as much as possible. There was an apprehension that the membership of a Western–sponsored and dominated consortium might restrict its freedom of manoeuvre.

The donors, however, insisted that without a formal consortium, as in the case of many developing countries receiving large amounts of assistance, it would be difficult to expand the quantity and improve the quality of foreign economic assistance which was so urgently needed by Bangladesh. It was suggested that the World Bank, as chairman of the consortium, could take the lead in promoting the case for greater economic assistance to Bangladesh by providing an independent technical evaluation of her economic prospects and performance. Faced with the acute balance-of-payments crisis in mid-1974, it was the government of Bangladesh which requested that the World Bank call an emergency meeting of donors to examine whether the quantum of aid flow could be increased and whether disbursements of already committed aid—commodity aid—could be accelerated. It was also at this stage that the government invited the World Bank to constitute a formal and a regular consortium of donors under its chairmanship, as had been suggested by the donors for so long; from now on the World Bank was to undertake annually a monitoring of Bangladesh's economic performance, as well as an assessment of its aggregate requirements of aid and its composition.

Bangladesh also decided in the same period to draw on the second credit tranche from the IMF. Unlike the first credit tranche, which was fairly unconditional, the drawings on second credit tranche of the IMF were subject to Bangladesh agreeing to meet specific conditions regarding her monetary and fiscal policies. Therefore, earlier in the year she was less than enthusiastic about drawings on the second credit tranche, but by the middle of 1974 she had no choice but to request drawings on it.

Fluctuations in imports had a direct impact on government revenues because about 40% of tax revenues and 30% of combined tax and non-tax revenues were derived from customs duties. In addition, a further 40% of tax revenues and 32% of combined revenues were derived from excise duties and sales taxes which were highly sensitive to the availability of imported materials and semi-finished goods. Essential imports like food and fertilizer either entered free or were lightly taxed, as was the case for cement and raw cotton. The greater the shortage of foreign exchange resources, the lower was the proportion of dutiable imports and the lower were receipts from import taxes and from sales taxes levied on imports. Domestic output of cigarettes and petroleum products based on imported materials was a major source of excise duties and their yields reflected fluctuations in the availability of imports. Another factor was that while the revenues of the government (including the surpluses of the nationalised sector) fluctuated, in absolute amounts as well as in terms of percentage shortfalls from the projections, current government expenditures in contrast, whether projected or realised, increased continuously. In 1972/73 actual receipts from custom duties were 40% below projections; in subsequent years, the divergence between projections and actual realisations was smaller. The first two annual budgets projected a surplus on revenue account but ended up with a deficit, although of diminishing magnitude over the years, which changed to a small surplus in 1974/75. Changes in the world prices of imports also had an impact on government revenues and on the extent of the budgetary surpluses or deficits. For commodities taxed at ad valorem rates a rise in the prices of imported goods could be expected to increase government revenues. A rise in the price of imported foodgrains did not have this effect, not only because imports were tax free but also because it was sold through the ration shops at less than cost and a rise in price was met by increasing the implicit subsidy to the consumer. Thus larger imports of foodgrains at higher prices paid for out of the government's own resources had the effect of reducing government revenues.

(d) Annual Plans and Price Changes

The implementation of the Annual Plan was marked by a significant change in absolute as well as in relative prices. The rapid and significant changes in both absolute and relative prices affected all monetary magnitudes—government revenues and expenditures, as well as cost estimates and rates of return on development projects. Furthermore, changes in prices made forecasts diverge increasingly from actual or *ex post* magnitudes. Given the deficit in both the balance of payments and the government budgets, even an equiproportional rise in prices, affecting both the debit and credit items in the balance of payments or government accounts, had the effect of widening the gap in absolute terms between payments and receipts.

Bangladesh suffered a dramatic change in both absolute and relative prices. The changes in relative and absolute price levels were already illustrated in the case of foodgrains, imports and changes in the terms of trade. The changes in foodgrain prices and in import and export prices were primarily due to factors over which Bangladesh had no control. While no adequate price indices were available, the cost of living index for the city of Dacca registered a fourfold increase between June 1970 and December 1974, implying an annual rate of inflation of about 75% per year. This probably understated the severity of inflation, since the index was largely confined to government employees and included controlled prices, rather than free market prices which were higher. During the same period the price of medium quality rice and of long cloth (coarse cloth for use by the masses) rose by more than 700%, that of sugar by more than 900%, while the prices of other staples such as soap, salt, firewood, rose between 300 and 1,000%. During 1972/73, the first year after independence, inflation was most severe—prices rose by 108% above the level of 1969/70; it moderated in the next year, i.e. about 39% compared to the previous year, but during the first half of 1974/75, following devastating floods, the rate of inflation again accelerated.[23]

A large number of factors including the rise in import costs, the shortage of imports and especially of raw materials and spare parts needed to expand domestic output, insufficient stocks of foodgrains, general supply constraints and inefficiencies in the distribution system all contributed to the inflationary pressure. The GDP at factor cost in constant prices of 1972/73 did not recover to the level of 1969/70 until 1975/76. By then the population was about 18–20% greater. Income and foodgrain consumption per capita remained persistently below 1969/70 levels. A rise in the price of rice was an important factor contributing to inflationary pressures; autonomous increases in wage costs were not an important contributory factor in an overall sense. The organised labour unions operated only in a small segment of the economy, and the effect of high wages on aggregate demand was not important, although the resulting rise in costs affected prices charged by the large-scale manufacturing industry. The rise in wages did not catch up with the rise in prices in the large-scale manufacturing sector. Outside the large-scale urban sector, wages did not respond to prices in any case. Skilled construction workers had enjoyed the greatest increase in money wages (about 60% increase in money wages during 1972/73); nevertheless they suffered a loss of real income of about one-third. Other workers such as skilled industrial workers suffered a much greater decline in real income.

The rise in import prices of 115% between 1972/73 and 1974/75, though important in other indirect ways, was not a direct major influence on internal prices since imports accounted for only 10% of the country's domestic product.[24] The direct effect on prices thus would have been no more than 12%. In 1974/75 when import prices rose by only some 5%, the domestic price level rose by 50%. Like the rise in wages, it did, however, increase the cost of domestically-produced manufactured goods which in turn affected the rate of inflation.

The shortages caused by stagnant domestic production and by the limited amount of imports, plagued the Bangladesh economy throughout all these years; while the cost-push factors due to rising import price and wage costs contributed its share to inflationary pressure, there was also a significant demand pull on the price movements exercised by the creation of bank credit. During the period between 1972/73 and 1974/75, money supply increased at an annual average rate of 35%. Deficit financing by the government and losses of the public sector enterprises were the prime causes of the credit expansion. Over the whole period the private sector contributed relatively little to the increase in bank credit. In 1972/73 it repaid debt and in 1974/75 borrowed little. Credit was needed in 1973/74 because import prices rose sharply and again in 1975/76 when import prices increased as a result of a devaluation of 58%, 30% of the expansion of bank credit went to the private sector. The slowing down in the rate of creation of bank credit, evident in 1974/75, might have continued in 1975/76, had there been no devaluation. The change in the magnititude and composition of domestic liquidity is seen on p. 152.[25]

The period was marked not only by inflationary pressures but also by significant shifts in relative prices. A comparison between wholesale prices of agricultural and industrial products, based admittedly on inadequate data, seems to indicate that there was a fall in the prices of agricultural products in relation to those of industrial products between 1969/70 and 1971/72. The relative prices obtained by the agricultural sector reached their lowest level at the end of 1972; by the end of 1973 there was a recovery, as may be seen in the table on p. 152.[26]

This implies that the general inflationary situation in the post-independence period was associated with adverse movements in

CHANGES IN DOMESTIC LIQUIDITY
(*million Takas*)

	1972/73	*1973/74*	*1974/75*	*1975/76*
Domestic Credit	*2,814*	*3,302*	*1,501*	*2,158.4*
Government	1,291	1,723	229	400.0
Other Public Sector	1,713	1,052	1,319	1,123.1
Private Sector	−190	527	47	635.3
Foreign Sector	*−188*	*−832*	*1,098*	*−1073.6*
Other	*273*	*−196*	*−1,888*	*250.0*
Total Domestic Liquidity	*2,899*	*2,274*	*712*	*1,334.8*
Money Supply	*2,103*	*1,207*	*−22*	*−803.0*
Time Deposits	*796*	*1,067*	*734*	*531.8*

INDICES OF WHOLESALE PRICES AT DACCA
(*1969/70 = 100*)

	1971 September Average	*1972 Sept. & Dec. Average*	*1973 Sept. & Dec. Average*	*1974 Sept. & Dec. Average*	*1975 March & June Average*	*1976 Sept & Dec. Average*
Agricultural Products:						
(a) General	117	172	217	442	417	331
(b) Food	112	175	247	556	513	370
(c) Raw Material	125	168	171	274	274	274
Industrial Products:						
(a) General	139	270	259	490	483	434
Terms of Trade (Agriculture/ Industry)	84	64	84	90	86	76
Terms of Trade (Food/Agricultural Raw Materials)	90	104	144	203	187	135

the terms of trade of the agricultural sector and, compared with the pre-independence period, with a shift in relative income away from the agricultural sector. However, from 1973/74 onwards there was some improvement. Within the agricultural sector, prices of food grains relative to other products were lower in 1971 than in

1969/70. From the middle of 1971, prices of foodgrains relative to those of agricultural raw materials rose to a level higher than that in 1969/70, and continued to improve until the middle of 1975. They declined at the end of 1975, but remained at a higher level than in 1969/70. This movement favoured the farmers and helped to offset the relatively higher prices of manufactures.

The rise in the price of rice was the most important component of inflation. The average retail price of medium quality rice rose from Tk.46 per maund in 1969/70 to Tk.108 in 1972/73, Tk.120 in 1973/74 and Tk.252 in 1974/75, coming down to Tk.174.4 in the first half of 1975/6.[27] Rice output constituted 40% of GDP. The movements in rice prices dominated the cost of living. The prices of many other commodities moved in sympathy with the price of rice. The peculiarities of rice marketing and distribution, the seasonality of its domestic production and the small proportion of output which was marketed all contributed to the volatility and inflationary pressure in the price of rice. In between harvesting seasons, demand for rice was met from stocks and imports. As we have seen during these periods, the price of rice was particularly vulnerable to inadequacies in the domestic stocks of foodgrains and also to delays in the arrival of the imported foodgrains. Since rice was a staple consumption item and minimum requirements had to be met, the shortfall in its supply was matched not by a fall in demand but by consumers selling liquid assets or incurring debts or reducing their consumption of other commodities. Outside the public distribution system, rice markets were dominated by small producers and the traders. When there was a shortfall in production, in imports or in the amounts distributed under the rationing system, fears were generated regarding future shortage and further rise in prices. Bottlenecks in transportation and inefficiencies in the administration of the public distribution system aggravated the effects of shortages on expectations. This encouraged the hoarding of foodgrains by traders and farmers, which accentuated the rising trend in price. As noted earlier, domestic stocks of foodgrains in the hands of the government were very low and could not be used to stabilise the market price of rice.

The upward movements in the price of rice coupled with the effects of the over-valuation of the exchange rate on the price of jute, resulted in a decline in the relative price of raw jute *vis-à-vis* rice. Between June 1972 and December 1974, the average price of jute increased by only 67% while the price of rice increased by nearly twice as much. Between 1969/70 and 1974/75 acreage under jute declined from 2.5 million to 1.4 million acres; the resultant scarcity and a rise in the price of raw jute increased the cost of manufactured jute goods and led to losses in the jute manufacturing industry. Thus the rise in the price of rice, by reducing the supply and raising the price of jute, affected the profitability of jute manufacturers.

The most significant impact on the annual planning exercises of the changes in absolute and relative price were the cost overruns on the development projects; it was necessary to revise the investment costs of most development projects. Since relative prices of inputs and outputs changed, the cost-benefit ratios and the choice of investment projects were also affected.

An important operational question in this connection was how to provide for rising prices in the estimation of the total costs of the Five-Year Plan, which had been formulated in constant prices. Since a rise in prices in most cases was almost inevitable would it not be desirable to take account of this explicitly in the cost estimates? There were good reasons why provision should be made for a revision of cost estimates of individual projects at the time of the formulation of the Five-Year Plan, as well as at the time of the initial evaluation of an investment project. In many cases all the elements of costs were not known. In a few cases, further studies were necessary and new facts needed to be collected; moreover there was the need for provision for inflationary rise in costs and prices. Since Annual Plans were formulated and financial requirements of investment projects were estimated every year in current prices, increasing the cost estimates of investment projects by a margin to take care of uncertainties and inflation, would have enabled a more realistic assessment of the resource requirements of the Five-Year Plan.

(e) Annual Plans: Organisation and Management

The Five-Year Plan and the Annual Plans laid considerable stress on the improvement of organisation and management as a key element in the fulfilment of the planned targets. It was the most important factor involved in getting the economy back to the level of operating efficiency reached in 1969/70.

One problem was securing co-ordination between the ministries and executive agencies both nationally and locally, particularly between the agricultural development agencies. Another problem was that of delegating authority from upper to lower strata in the hierarchy of the government. This related to the delegation of authority from the minister to his officials in the ministry as well as from the officials in the ministry to the autonomous agencies or departments in charge of implementation of projects or of the operation of an enterprise. It also involved the specifications of the rights and obligations of the agencies or officials at the various stages of execution or at the various levels of government; it required the recognition of the principle that delegation of responsibility along with an appropriate system of accountability was the most effective method of administration. There was also the problem of giving adequate incentives, or instituting a system of reward and punishment, relating to the performance, efficiency and motivation of the functionaries operating at various levels of the government.

Both the Five-Year Plan and the Annual Plans spelled out in great detail the problems of co-ordination, delegation of authority and motivation or incentives for executives and officials along with recommended approaches towards the solution of these problems in the various sectors of the economy. Improvements were slow to occur and the effect of slow improvement was felt in continued deficiencies in the field of organisation and management of the economy. In the sphere of implementation of development projects there were a number of critical deficiencies, some of which could be identified as follows:

(1) The cumbersome procedures for the approval of projects, for making decisions for procurement as well as for the approval of contracts for consultants or for construction companies or agencies, and for the appointment of staff for the projects themselves.
(2) Lack of forward planning for the project inputs, caused partly by lack of funds and partly by administrative procedures.
(3) Lack of inter-ministerial or inter-agency co-ordination.
(4) Delays in the release of funds by the Ministry of Finance and clearance of imports through the customs. One of the factors causing delay was the uncertainty of the estimates of revenue and the consequent policy of the Ministry of Finance to maintain tight control over development expenditures. There were, in addition, further complications and red tape involved in the release of funds from the Ministry of Finance to the executive agency via the intermediary of the ministry which controlled the executive agency itself.
(5) Delays in deciding how to deal with cost-overruns resulting from inflation, which were rampant during this period.
(6) Promotion and salary policies which discouraged initiative and decision-making and which rewarded loyalty, seniority and aversion to risks.

The most important aspect of the centralisation of administrative and financial powers involved the lack of sufficient authority on the part of the government agencies and ministries to make purchases or procurements for the projects or enterprises under their control. All procurement decisions and contracts beyond Tk.1 million or $140,000 had to be approved by the Prime Minister and later on by the President when the parliamentary government was changed to the presidential system. Prior to January 1975 there was a Cabinet sub-committee consisting of ministers of finance, trade and industry which was charged with scrutinising all bids (i.e. all offers to sell goods) and then making recommendations to the Prime Minister for his approval. There were several stages in the process. Firstly, documents relating to alternative bids received by the agency concerned had to be submitted to the ministry (Secretary and the Minister) under which the agency was located. Moreover, there were two stages in the examination of bids: one was at the level of the executive agency and the other was at the level of the ministry, before the "bids" were passed on to the ministry for scrutiny. This was followed by a cross-examination of the officials of the ministry and agency by the Cabinet sub-committee before it made its final recommendations to the Prime Minister. These various stages of examination led to a great deal of paper work and documentation and consequent delays in the approval of purchase and procurement contracts.

The purpose of the centralisation of the decision-making process was to reduce the possibilities of corruption and graft on the part of officials in the agencies and in the ministries by having the Cabinet sub-committee and the Prime Minister vet the procurement contracts. In order to reduce delays in the process of decision-making caused by the possibility that cabinet ministers might not always be readily available, arrangements were made for the Prime Minister to act on behalf of the ministers when decisions were needed urgently. Delays occurred not so much because the Cabinet sub-committee did not find time to meet at short notice or took a long time to make decisions but because a considerable amount of time was taken preparing additional documentation for the presentation of the case to the Cabinet sub-committee, since the committee seldom had any foreknowledge of the details of the particular contracts. This would have been unnecessary and much time could have been saved if the agencies themselves were empowered to take decisions.

The involvement of the Cabinet sub-committee in the procurement decisions was a departure from the pre-independence practice, when the agencies had greater financial autonomy. The institution of the Cabinet sub-committee, in so far as the officials were concerned, reflected a lack of sense of trust or confidence in them. This created a sense of disaffection which militated against their efficiency. Moreover, the involvement of the Cabinet sub-committee in administrative matters of such detailed nature detracted from the time and attention they could devote to policy matters. It was doubtful whether the cost of this procedure in terms of delays and use of the scarce time of the Cabinet ministers was not greater than any benefit from a possible decline in graft and corruption on the part of officials. That politicians were always less liable to corruption than such officials was debatable and in any case there was nothing to prevent a coalition between the ministers and officials in corruption. In fact, there was a widespread belief among the community at large, in the prevalence of significant corruption and misuse of power for patronage on the part of the politicians. Whether in fact there was widespread corruption or not was less important than the fact that its existence was widely believed. This detracted seriously from the credibility of the government in the eyes of the public. In the atmosphere of such a widely held belief in the lack of integrity of government, it was difficult to muster support for its policies or to generate the conviction that they were not conceived in the personal interests of individuals.

There were twenty or more ministries and often closely related activities or functions were split up between the different ministries. The problem of inter-ministerial co-ordination in the execution of functions was nowhere more acute than in the field of agriculture and transport. The functions of agricultural development were split between the three ministries, i.e. those of Agriculture, Water and Flood Control, and Rural Development. The procurement, supply and distribution of agricultural inputs such as fertilizers and seeds was exclusively in the hands of the Ministry of Agriculture whereas the provision of irrigation facilities was the joint responsibility of the Ministry of Agriculture and the Ministry of Water and Flood Control. In addition, there were small irrigation and drainage schemes which were under the jurisdiction of the Ministry of Rural Development. These small-scale schemes were to be administered under the rural public works programme, which was the responsibility of the Ministry of Rural Development. The extension services were split between the Ministries of Agriculture and of Rural Development. The former provided the services of agricultural extension workers up to the union and village levels; the latter organised rural co-operatives and provided extension services through "model farmers" and managers of co-operatives which it trained. The basic objective of the Plan was gradually to cover the entire country with the village level co-operatives which would replace the extension services provided by the village or the union-level extension workers. However, the coverage of the village level co-operatives, which integrated the function of extension, credit and supply of imports, was as yet limited. The major part of the country still had to be covered by the extension workers of the Ministry of Agriculture. Moreover, the training of the "model farmers" in the co-operatives in the areas covered by the cooperatives, was undertaken by the specialists and extension workers of the Ministry of Agriculture. The supply of credit to the farmers was partly under the control of the co-operatives, and partly under the control of the agricultural banks, which were under the control of the Ministry of Finance. The Ministry of Agriculture felt that since it was in charge of increasing agricultural production, the supply of all inputs including credit, irrigation and extension services should be placed under its unified command in the interest of efficiency. On the other hand, it was argued by those in favour of the status quo that control and regulation of all functions under one ministry would be too unwieldy and inefficient and would generate diseconomies of scale.

The Five-Year Plan suggested that there should be a system of co-ordination by committees at all levels: one at the level of ministers for the purpose of policy-making, and another at the level of officials of the three ministries for the purposes of co-ordination of administrative or executive functions. In the thanas and unions (local government units) co-ordination was to be achieved by the local officers of the various ministries meeting under the leadership of the elected local representatives whenever there were elected local governments, or under the local administrative officer. The problem of effective co-ordination between the three ministries, nationally as well as locally, remained unresolved throughout this period. In fact, the co-ordination committee advocated at the national level was not constituted until 1975/76.

There was also the problem of co-ordination between the Ministries of Agriculture and of Jute for the responsibility of implementing the development programme for jute production. The Jute Ministry which regulated the jute manufacturing industry, jute trading, both internal and external marketing as well as technological and scientific research on raw jute and jute manufacture, advocated bringing the responsibility of promoting efficient cultivation of raw jute under its control, on the ground that it was an integral part of its responsibility of increasing exports and supplying raw material for the manufacturing industry. The Ministry of Agriculture on the other hand, contested that the functions of providing and stimulating the use of modern inputs, credit and extension services to the jute growers were not different from those in respect of rice growers. In fact, jute and rice were competing crops, involving alternative uses of land and had a close interrelationship in terms of development problems. The solution suggested by the Five-Year Plan was that it should remain the responsibility of the agricultural ministry but an inter-ministerial committee should co-ordinate policy and the execution of the jute development programme. The controversy and argument dragged on for more than two years and the uncertainty about the division of responsibilities between the two ministries caused delay in the execution of the programme.

The most outstanding consequence of inadequate managerial and organisational inefficiency, on the one hand, and lack of trained and supervisory personnel, on the other, was the widespread prevalence of under-utilised capacity in the various sectors of the economy. A few selected examples from the field of agriculture, industry, transport and electric power would illustrate the nature and consequences of the general problem mentioned here.

The shortfall in the implementation of the irrigation development programme, in terms of delays in the sinking of tubewells and low-lift pumps as well as in the completion of the multi-purpose irrigation project has already been mentioned. What was equally crucial was the under-utilisation of capacity of the irrigation facilities already established. This illustrates how critical were institutional and pricing policies for the proper utilisation of installed capacity and how in the absence of appropriate organisational framework and managerial efficiency, resources already invested might go to waste in a country which was already acutely short of capital resources.

The "command area" or the amount of land effectively irrigated by low-lift pumps (surface-water irrigation pumps) and by tubewells (ground-water irrigation wells) was not only much below their optimum capacity but also was declining over the years. Nowhere was the actual command area more than 50% of the optimum command area and it was often nearer 25% than 50%. The reasons for the poor performance of the irrigation pumps were identified as follows:

(1) Placement or "fielding" of pumps without a careful analysis of the availability of water in the surface-water channels, rivers and streams. In many cases where pumps were placed, streams or river channels dried up before the winter crop was sown. Frequently, they were placed without proper spacing, closer than was warranted by the availability of water. This was often done in response to the pressure of local influential farmers, who wanted pumps nearer their lands, even though the availability of water did not justify it.

(2) It was not only that pumps were placed without an appropriate assessment of the availability of water, but also they were not placed in the right spot on the river embankment. Frequently a low-lift pump was placed on the highest spot on the river embankment. The greater the lift a pump was used for, the greater the consumption of fuel and wear and tear of the pump.

(3) The government agency in charge of the procurement of pumps sometimes procured inappropriate pumps which required more fuel than others of the same discharge capacity.

(4) The system of organisation of village pump groups, introduced

before independence, was ineffectively enforced. A pump was much too big for one farmer and pumps were to be provided only to a minimum group of forty-five farmers. The local government officials did not strictly enforce this condition. Both the local government institutions and local offices of the central agencies were involved in the programme and there was ineffective co-ordination. The entire process of selection of sites for the placement of pumps, organisation of groups of farmers, and training of the farmers to make the best use of water was the responsibility of the various government agencies and the co-operatives. Wherever there were co-operatives, they did not discharge their responsibilities efficiently. Altogether there was inadequate training of farmers and inadequate extension services.

(5) The distribution of water from the site of the pump was wasteful and inefficient. There were inadequate or inefficient feeder channels. The construction of the feeder channels which were to distribute water to the plots of the individual farmers, was left to the responsibility of the farmers. The farmers lacked technical skills in construction of channels or in water management. The rural public works programme did not provide for the expenditure of public funds on the construction of channels. Moreover, in some areas, the levelling of land of all the farmers included in the pump group, which was essential for prevention of wastage of water, did not take place.

(6) There was not only an inadequate supply but also a maldistribution of fuel and spare parts among the pump groups. The distribution of fuel by the government agency (Bangladesh Agricultural Development Corporation) which was in charge of fielding and maintaining the pumps was not strictly related to the number of hours individual pumps worked. No systematic and reliable records were kept of the working hours of the individual pumps. There was no local manufacture of spare parts; imports of spare parts were erratic and their distribution among farmers was haphazard and irregular. The miscallocation of fuel without regard to the actual requirements of pumps resulted in some owners of pumps having excess fuel and others having less than their requirements. The highest number of hours a low lift pump was worked was about 5.5 hours a day even though a pump could be used 16 hours a day to meet peak irrigation requirements.

(7) The pricing of fuel was below its market price with the result that there was excess demand for fuel which was diverted to other uses than pump operation. The optimum utilisation of irrigation water depended on the availability of complementary inputs such as fertilizer and high-yielding variety of seeds. Failure to synchronise the supply of water and complementary inputs resulted in adequate use of pumps.

(8) The major disincentive in the way of efficient use of pumps was the very low rental which was charged for them and the considerable subsidy which was implied in the system of rentals. The high profits derived from the use of pumps left no incentive for their most efficient use. An evaluation study reported that in one area 60% of the farmers were prepared to use a pump if the subsidy was completely withdrawn. In addition to the large subsidy on the use of pumps there was little financial discipline in the supply of credit by agricultural credit institutions and co-operatives, which recorded large defaults and delays in repayment.

(9) The "pump groups", the formation of which was obligatory to secure an entitlement to pumps, were very loosely constituted groups—without any intention of co-operation in the general field of agricultural operations, apart from getting access to the use of pumps. They were not motivated by any sense of economic purpose, collective responsibility or mutual vigilance or control. They did not have the characteristics of co-operatives, though it was expected that having been brought together for the purpose of getting access to the use of water, they would also co-operate in respect of the agricultural operations which would help to obtain the maximum benefits out of the use of pumps. This might have been encouraged but for the fact that large subsidies on the use of water were available. The pump groups were often formed without strict attention being paid to the proximity of the various pieces of land held by the different members of the groups. The members with land farther away from the pump could not get adequate water, and even if they did there was considerable seepage and wastage of water by the time the flow reached the more distant plots. The nearer plots were over-supplied with water whereas only a bare minimum quantity of water was left for the more distant farmers.

(10) The responsibility for operating as well as maintaining the pumps was retained by the government and not by the user or beneficiaries. The inefficiencies were acceptable to the user because the cost of water was so nominal. To make matters worse, even the nominal rent was not, and could not, be collected regularly.

(11) The manager of the pump groups, in fact, acted not as a leader of a co-operative but as a middleman, in many cases paying rentals and price for fuel to the government in advance of his collecting them from the farmers. The composition of the pump group and the manager were frequently changed from year to year. The shifting composition of the pump groups was partly due to the fact that sharecroppers for the particular prices of land changed from year to year. Improvements in land or water management for the efficient use of water were naturally discouraged by the lack of a permanent stake or interest in the plots of land covered by the pump groups. The pumps wee returned every year to the government for repair,

maintenance and check-up and were taken back by the farmers after the payment of rentals. With a shifting composition of the pump groups, the original manager found it impossible to collect his loans.

The foregoing analysis illustrates how organisational deficiencies and inappropriate pricing and subsidy policies led to the inefficient use of capital and equipment and to low yields on the investment. The Annual Plan of 1973/74 offered a number of concrete suggestions for the improvement of the operating efficiency of pumps. For example, surface water potential in each thana was to be estimated to provide the basis for determination of the appropriate number of pumps to be installed in each thana. The technical considerations, rather than a response to the demand of pressure groups based on the supply and demand for water in each area, were to determine the location of pumps or the distribution of pumps between competing areas. It was necessary to define precisely the responsibility and improve the efficiency of the Bangladesh Agricultural Development Corporation (BADC) in respect of technical and supervisory services required for the design and construction of irrigation channels and for the installation of pumps and wells. Moreover, while the overhaul and repair facilities should be provided by the BADC, operation and routine maintenance of pumps should be the responsibility of the peasants' groups trained by the BADC. The hours of operation should be more than doubled; farmers should be motivated accordingly and should be provided with efficient extension and training services. Rentals for irrigation quipment should gradually be increased so as to cover the entire costs.

A few improvements in policy were introduced later in 1974 when the rental for the pumps was increased and responsibility for training. controlling and paying the pump operator and for continued repair and maintenance was transferred to the pump groups. The government was to provide the spare parts and fuel at market price and the responsibility for purchasing them from the government remained with the farmer's groups. Moreover, the government undertook to provide training to the drivers and mechanics appointed by the farmer's groups for running and maintaining the pumps. An innovation was the provision for the purchase of the pumps in instalments by the farmer's groups operating them.

The Bangladesh Water Development Board (BWDB) was in charge of large-scale flood control and irrigation projects. The Plan, similarly, had specific recommendations for the speedy implementation of physical targets in this field as well. It identified the types of surveys and feasibility studies, including hydrological, ground water, aerial and hydrographic surveys as well as specific studies and hydraulic research on barrages and rivers. It suggested a time frame for the quickest completion of the small-scale, short-gestation irrigation projects. It recommended working out a timetable for the annual work programme of BWDB so that construction could be completed or a major part of it done in the winter season, since no construction could be carried out in monsoon seasons, and the rest of the preparatory work prior to the construction work could be completed before the winter months. It made recommendations regarding speedy release of funds for both domestic and foreign exchange expenditures as well as regarding the procurement of land and imported machinery and equipment for new projects. Also included were suggestions for training of workers and for repairs and maintenance.

The Ganges–Kobadak project was a multiple-purpose irrigation project which was started in 1954/55 and was completed in 1969/70. The total investment in the project was about Tk.230 million. It was notorious as a very costly project with an extremely inefficient utilisation of water. The project was designed in two phases. The first phase of the project was to be completed by 1975/76 and was to cost about Tk.380 million. Both the phases of the project were together expected to irrigate about 300,000 acres. But 50% of the water supplied by the project was wasted. This was due partly to technical defects in the project and partly to deficiencies in the institutional arrangements and policies which had not succeeded in training farmers and providing incentives for the optimum use of water. The total availability of water from the project could have been increased by 20%, but this was not done because whatever was available was not used. The main channel was not dredged nor were the flow and path of the river properly controlled. The technical arrangements relating to the pumping facilities required improvement. There was a shortage of water in high lands and excessive water and flooding in low-lying areas. The main causes of wastage of water were disruptions and damages to canal banks, failure to provide land channels, land levelling and an irrigation rotation system, and lack of extension work to train and stimulate farmers in the efficient use of water. In ten years no more than 20% of the area that could be irrigated had been levelled and provided with channels and instruction for farmers. The water was supplied free, and consequently the farmers had no incentive to make the maximum use of water. The collection of water rates was long overdue in spite of the fact that a specialised staff was attached to the project for this purpose. Seven agencies or government departments were involved in the project. The co-ordination between them was bad. There was no leadership at the project level guiding and co-ordinating the various agencies. The lack of co-ordination was partly responsible for slow progress in extension work and in securing the co-operation and increasing the competence of the farmers in the use of water. The protection of the embankment and associated work relating to the project were intended to be the joint responsibility of the farmers and the project staff.

The manufacturing sector in Bangladesh illustrated yet another instance of under-utilisation and inefficient use of capital stock and low yields from investments. In most of the industries the rate of capacity utilisation was lower than what it had been in 1969/70. The under-utilisation of capacity in the industrial sector was due partly to interruptions in the availability of power and shortage of spare parts and raw materials; but it was also caused by inadequate supervision and management, indiscipline among the workers and the absence of appropriate incentive systems for workers and managers.

Apart from under-utilisation of capacity, there were unexploited opportunities for increasing output per man, through small, low-cost improvements in the technical and organisational set-up. An evaluation report on the state of the biggest jute mill in the country reported that the mill could turn its current losses to a profit of about 18% of the sales value through improvements in organisation and productivity.[28] The measures were mainly to be directed towards increasing the volume of output, improving the use of raw jute and reducing the employment of excess labour, which was recruited during the war and in the post-liberation period. Improvements could be effected in the selection of appropriate qualities of jute and storing it properly to prevent deterioration, It was necessary that batches of raw jute of different qualities should be combined in appropriate proportions in order to improve the quality of jute yarn in terms of durability and moisture control; application of oil and moisture had to be done properly. The performance of the machinery was very low; spinning efficiency was about half of the "normal" level. There were no systematic machine-work records indicating programmes for inspection and cleaning of the specific parts of the machinery. Nor was there an organised overall routine for machine cleaning and lubrication. The mechanical maintenance of the jute mills organisation was at a low level of efficiency; the workshops needed overhaul. Different types of machinery were not available in the correct proportions and additional 'balancing' machinery was needed to attain this. The deficiencies in the system of financial control included lack of reliable records or data regarding production, costs, sales, inventories of raw materials and finished goods, etc. There was an absence of properly trained accountants, managers and supervisors conversant with the techniques of financial control and personnel management. With respect to personnel management and policies the improvements required were identified as follows: (*a*) improved recruitment, selection and allocation; (*b*) massive training programme for operatives, supervisors and management; (*c*) job description, clearly defining the responsibility and authority of every job; (*d*) increase in the number of labour officers for improving relations between workers and management; and (*e*) improved procedures for handling grievances and other human problems.

As in the case of the jute manufacturing industry so also in the second most important industry, there were considerable inefficiencies in both spinning and weaving due not only to the shortage of spare parts and components but also to inadequate repair and maintenance and lack of "process control".[29] The use of inappropriate raw material reduced the quality of output and led to a decline in the efficiency of the spinning process. A textile machinery engineering mission which investigated the cotton textile industry operative efficiency advised the organisation of a few efficient units, reducing the number of existing ones by using such machines as were serviceable after repairs or modifications. It suggested the establishment of a model plant in the industry as a centre for training skilled personnel as well as the establishment of textile machinery repair facilities. Within the industry itself, however, specialisation of plants in such processes as spinning, weaving, dyeing and processing, etc., were needed to increase efficiency. The low level of production in cotton textiles was due mainly to operative inefficiencies but also to such factors as strikes, lockouts, power failures, which were often outside the control of the industry itself. A study of the cotton textile mills during 1973/74 revealed that during the first nine months of 1973/74 about 6% to 7% of the output of yarn and cloth was lost due to four factors as shown below.[30]

LOSS OF PRODUCTION IN THE TEXTILE INDUSTRY IN 1973/74

	Percentage Distribution of Loss of Output	
Reasons for Production Loss	*Yarn*	*Cloth*
Power Failure	40	23
Lack of Spares	32	35
Strikes and Lockouts	11	21
Machine Breakdown	17	21

Yet another example of the inefficiencies in the industrial sector was the case of the only steel mill in Bangladesh which was designed originally on a scale less than that of an optimum size plant; moreover, it never reached full capacity operation. There was a lack of balance in equipment and machinery as well as between the blast furnace, steel plates, and steel billets plants. The three plants related to different vintages in technological development in steel making. There were serious deficiencies in the management of the factory, including defective accounting and inventory control procedures. One of the serious drawbacks was bad programming of imports and of the domestic procurement of raw materials which prevented an efficient rate of operation in the factory. There was no arrangement for the procurement of scrap from the domestic market, principally from the waste materials in the transportation and communication sectors. Though a very large part of requirements of scrap could have been met from domestic sources, the mill depended mainly on imports for the supply of both ingots and scrap; this not only involved expenditure in foreign exchange which could otherwise have been spared but, more important, irregularities or uncertainties of the supply of imports often led to interruptions in production which could have been partly avoided if reliance had been placed on the domestic supplies. There was also no arrangement for recycling rejects from the steel mill, along with ingot and scrap, into the production process.

There was again the case of the triple super-phosphate factory (fertilizer) which operated at a level considerably below full capacity output. It was a "turnkey" project built by the Japanese and there was no contract with the Japanese enterprise to provide management in the initial years and to train local personnel on the job in the factory. The factory was operating at 30% of its capacity in 1974; irregular supply of imported raw materials, inadequate protection of the machinery and equipment against sulphuric acid, incompetence of the technicians in operating the heating equipment and lack of quality control owing to inadequate technical training of the labour force, all contributed to inefficiency. The original factory was built on the assumption that it would be part of a complex, consisting of two additional plants, producing ammonia sulphate and white cement. Both these products were to be based on the utilisation of gypsum, a by-product of the fertilizer factory. But the original complex did not come into being, partly because of the disruption of the market after the break-up of Pakistan. While a new study was commissioned to decide on the most optimum use of the gypsum, the cost of fertilizer was increased by the substantial expenditure incurred in disposing of the continuously accumulating stock of gypsum.

Defects, managerial and technical, in physical infrastructure, especially in transport and communications, were a serious bottleneck to the recovery of the economy to its pre-independence level of operational efficiency. The different branches of the transportation network in Bangladesh were divided between the Ministry of Communications responsible for railways, roads, highways and ports, and an independent ministry regulating inland water transport, shipping and civil aviation. Under normal circumstances there was a need for co-ordination between the different modes of transport; they were competitive as well as supplementary. In post-independence Bangladesh it was vital to make the best use of all the available transport facilities. The damage caused by the war and its aftermath fell heavily on transportation. A large number of railway and road bridges were destroyed. In a country criss-crossed with waterways this caused great dislocation and disruption in road and railway transportation. It was not until 1974 that repairs and reconstruction were substantially completed in the transportation network. The position was exacerbated by the need to move large quantities of imported foodgrains. The UN relief operation which started immediately after independence, concentrated particularly on the transportation and distribution of food. Before 1972, railways carried more than 60% of the cargo from the principal port of Chittagong to the interior of the country. Because of the destruction of the railway bridges, water transport had to be used much more than before independence for the transportation of foodgrains from the port to the internal marketing and storage centres. The opening up of trade links with India, after many years of disruption of trade, involved redirecting the transport towards the western region, and required a balanced combination of water, road and water transport to the borders of India. In short, it was a very urgent task to secure very effective co-ordination of the different modes of transport and the maximum use of the diminished supply of transport equipment.

The Annual Plans in 1972/73 and in subsequent years greatly emphasized the need for appropriate institutional arrangement to co-ordinate the transportation system. To illustrate, the timing of the arrival of the barges and boats from the inland shipping points to the port of Chittagong had to coincide with the arrival of ships from abroad carrying imports, especially foodgrains. Again, they had to convey exportables, especially jute, from the interior of the country to the ports in time for the ships calling at the ports to load the exports for destinations overseas. Often boats and barges carrying jute to the ports had to remain at the port waiting for ships whose arrival could not coincide with theirs. The boats, in fact, served as floating godowns, with the result that the transportation needs of other commodities suffered. The problem of co-ordination between railways, ships and road transport was particularly difficult for foreign trade because while the shipping was under one ministry, the administration of the ports was under another. The

Annual Plan made recommendations for the establishment of a transport co-ordination division in the Ministry of Communications headed by a civil servant of the highest rank.

In addition, it was suggested that there should be regional offices located at the two ports for the effective co-ordination of all modes of transport. In fact, the recommendations of the Planning Commission for the formation of a separate division for transport co-ordination was not implemented; instead an inter-ministerial committee was set up headed by one of the ministers with a small secretarial staff, which had neither the status and authority nor the strength and competence to undertake the task which was expected of it by the transport administration. The Annual Plan of 1974/75 pointed out that the current structure of transport administration militated against the development of co-ordinated longer-term policies. The sectoral interests of the separate transportation agencies held sway and matters were made worse by the internal split of the Ministry of Communications into divisions with two apparently equal secretaries, one dealing with railways and the other with the rest of the responsibilities of the transport ministry. The Plan recommended the creation of a single ministry of transport.

A Bangladesh Transport Survey was undertaken by a foreign firm of consultants beginning in 1972, which among other things made recommendations for improvements in the operation, planning, organisation, administration and management of each transport mode, and suggestions for improvements in the government's transport policies, particularly for transport co-ordination. The Annual Plan of 1974/75 urged serious examination of these recommendations by the government, with a view to their adoption. An obstacle to improvement in efficiency was the lack of clearly defined powers and responsibilities of the ministries, on the one hand, and the various operating agencies, on the other. The power of the transport agencies in charge of running and operating the different modes of transport was unduly restricted. This was true of their financial as well as their administrative powers including powers of appointment and personnel policy,[31]

The organisational deficiencies and inadequate co-ordination partly contributed to and were partly aggravated by the existence of under-utilised capacity in the transport sector. In 1975 only 70% of railway engines and 56% of railway carriages were in operating condition. The rest of the engines and carriages were either damaged or waiting for repair. There were long delays in the schedule of repair and maintenance of carriages and engines.[32] In the field of road transport, less than 50% of the total trucks in the public sector transport corporation were roadworthy. The considerable lag in the schedule of repairs and maintenance was

SCHEDULE OF REPAIR AND MAINTENANCE
(*numbers*)

	1969/70		*1973/74*		*1974/75*	
	Target	*Actual*	*Target*	*Actual*	*Target*	*Actual*
Engines	166	145	180	88	180	69
Passenger Carriages	1,957	1,466	1,993	780	2,489	610
Freight Wagons	5,760	4,592	5,760	2,274	6,553	1,243

most noticeable in the railway sector as seen above.[33] The decline in the state of the repair and maintenance services in the post-independence period compared with 1969/70 is evident.

Along with the transportation sector, another component of physical infrastructure which suffered from operating inefficiency was the generation, transmission and distribution of electric power. The Annual Plan of 1974/75 identified the following principal reasons for the frequent interruption of power supplies as well as general inefficiency and low rates of return on investment in electric power.[34]

(1) There was a divergency of about 25% between the total amount of energy generated and the amount of energy recorded as sold. The difference was due partly to losses in transformation, transmission and distribution and partly to pilferage, inaccurate metering, incorrect meter reading and billing. With appropriate corrective measures losses could have been cut to 15% from the high level of 25%.
(2) The irregular supply of fuel oil, from both imports and domestic production of the oil refinery, and delays in the transportation of fuel to the power-generating plants resulted in interruption in the generation of power.
(3) Thre was a dearth of properly skilled manpower and excess employment of unskilled labour. In addition, there was the reluctance of the workers to move from plants with excess labour to plants short of labour.
(4) There were inadequate facilities for repair and maintenance.
(5) The accounting and auditing system was faulty. There was no reliable accounting of assets and liabilities so that the proper depreciation rate and adequacy of depreciation reserves could not be determined.
(6) There was no adequate inventory control and programming, nor timely and appropriate financial, operating and progress reports.

The above analysis illustrates how widespread were the deficiencies in the matter of management and organisation and how important was their impact on output and income. They also point out the opportunities which existed for improving the efficiency with which capital stock in almost all sectors of the economy could be used and how, with appropriate changes in training, skills, methods and habits of work, organisation and institutions, gains in productivity could be obtained. Given the acute scarcity of resources in Bangladesh, the need for greater efficiency in the use of resources was most urgent. However, it was paradoxical that there was frequently great pressure for adding to capital stock when existing stock was either under-utilised or inefficiently used. It was often a soft option on the part of the agencies and ministries, which were insufficiently aware of the scarcity of resources and its long-term adverse consequences for all of them.

(f) Revision of the Five-Year Plan

The experience with the Annual Plans shows how administrative deficiencies and unforeseen events invalidated the assumptions made in the Plans. The recovery of the operating efficiency of the economy to the pre-independence level was much slower than was expected. The improvements in organisation and management and changes in the institutional framework, which were required for a speedy recovery from the damages and dislocations of the war, were found to be a good deal more difficult to achieve than was originally anticipated. This has been amply analysed in the foregoing pages. Implementation of the development projects included in the Plan was slower than expected. This also was partly due to managerial or organisational deficiencies, but shortage of resources was also a factor. Both foreign exchange and domestic resources fell far short of the projections. The successive crop failures, droughts and floods accentuated the state of stagnation in agriculture. The world-wide inflation and deterioration in the terms of trade further constricted the availability of resources for development. It seemed that, as had already been pointed out earlier, the projection of available resources was over optimistic.

The actual rate of saving in 1973/74 was much lower than expected in the Annual Plan, notably because public saving was negative. One estimate put domestic saving at 1.5% in 1973/74 and had forecast that it could rise to 4% by 1977/78—this would imply an average rate of saving of less than 3% during the Plan period.[35] Another estimate did not even anticipate a 3% rate of saving to be reached before 1980.[36] The post-independence experience in mobilising domestic resources emphasised the difficulties and the enormity of the task involved in the fulfilment of the Plan targets. The low rate of domestic savings was partly due to the continued stagnation in per capita income, especially in agriculture, with food shortages and rising inflation cutting into the margin for savings. Past estimates of the savings–income ratio varied between 8% to 10% during 1965/70, with considerable year-to-year fluctuations; the savings-income ratio was highly sensitive to changes in per capita income. Per capita income in 1972/73 was between 15% and 20% below the pre-independence level.

The estimate of the expected net disbursement of external resources of about $2.3 billion anticipated in the Plan was considered very optimistic at that time, but in practice a much higher level of assistance had been available to deal with extraordinary balance-of-payments difficulties. Actual annual disbursement of foreign economic assistance by 1975/76 exceeded $1 billion. Whether it was likely to continue on this scale in future years was a matter of judgement. Even if this proved to be the case, expected total resources, available from both foreign assistance and domestic sources, were likely to be insufficient in real terms to achieve the targets of the Plan. The terms of trade had deteriorated drastically. Export expansion fell considerably short of that intended. Moreover, the time lag required in the acquisition of skills and managerial capacity seemed to have been underestimated.

Faced with a changed forecast of available resources and a new set of relative prices and costs, one alternative would have been to adjust the Five-Year Plan within the context of the Annual Plans. This was done during 1973/74 and 1974/75. However, the need for a new framework of a long-term perspective was soon felt. During the Annual Plan periods, decisions had to be taken on new projects in each sector in the context of an evolving pattern of future demand and related developments in the other sectors. While it was true that in the first stage or as a preliminary step, each sector had to re-examine its intrasectoral priorities in the light of new prices and costs, the relative pattern of import substitution in different sectors was also to undergo changes. This could not be done within the confines of an Annual Plan. A long-term plan, for a particular sector, which was not linked together with other sectors within an overall framework, could not be formulated. For example, investments in physical infrastructure were to undergo substantial alterations, in response to changes in the rate of growth in agriculture and manufacturing industry.

The revision of the Five-Year Plan became essential also for a very practical reason. The ministries and executive agencies became very concerned with retaining and securing the sectoral allocations at least up to the level provided in the Plan—all the more as they were faced with the possibility of a curtailment in their programmes. At the time of the preparation of the Annual Plans, the original sectoral allocations of the Five-Year Plan were invoked by the ministries as a basis of their respective claims on total resources. It was not sufficient to work out an implicit set of priorities and criteria, without actually preparing a new Plan which could be used at the time of the Annual Plan to screen and reduce the sectoral programmes and the projects. Without a new framework, which was clearly understood by all the agencies and ministries, a readjustment of the priorities and inter-ministerial allocations proved to be difficult at the time of making Annual Plans; there was every possibility of distortion in the allocation of resources if the ministries did not appreciate the changed priorities and did not respond to the new situation. Under normal circumstances, the Annual Plan was intended to accomplish these sorts of adjustment in the light of new and changing situations. However, since the changes in resource availability and relative prices were very substantial, the original Five-Year Plan could no longer serve as a framework of reference.

One simple way of compensating for the shortfall in resources was to increase the number of years over which the Plan was to be implemented. In other words, the development programme of the Five-Year Plan would be implemented, let us say, in eight years or in ten years. All the projects and programmes would remain as they were in the original Plan. This, however, did not eliminate the role of choice; since all projects could not be undertaken at the same time, the timing of projects became relevant. The projects selected for implementation during the next five-year period or during the successive Plan periods had to be consistent with each other in terms of inter-sectoral relations as well as in terms of meeting the projected level and composition of final demand at different points of time. Moreover, this particular way of revising the Plan could not take into account the new production possibilities opened up in view of changes in prices and costs. Any revision of the Plan, therefore, involved two sorts of adjustment: postponement or staggering of some of the projects which were already included in the original Plan and, abandonment or modification of existing projects or introduction of new projects based on new sets of comparative costs or prices and new technological and economic data.

Towards the end of 1974 tentative steps were taken to re-examine the Plan. The first step was the estimation of costs of projects and programmes in each sector in terms of the new prices of 1974/75. In 1974/75 prices the financial outlay required to implement the physical content of the first Five-Year Plan was 80%–100% higher than the original estimates based on 1972/73 prices. The next approach in revising the Plan was to estimate the extent of reduction in physical programmes of the various sectors which would be necessary if financial allocations were kept unchanged. Since resources were expected to be less than originally anticipated, the next step was to conceive of a Plan which would be smaller than the original in both financial and physical terms. This could not be done by imposing proportionate cuts on all sectors; there was an inescapable need for readjustment of priorities and sectoral allocations. Thoughts, therefore, turned towards formulating a hard-core programme. The main criteria in formulating such a hard-core programme were to assign priority to quick-yielding projects and projects which were in advanced stages of completion; agriculture, especially the production programmes for rice and jute, was to occupy a central place in such a programme. Short-term gains in foodgrain production were high and comparative advantage for Bangladesh was substantial. Within such an order of priorities, an attempt was to be made to preserve inter-sectoral consistency. The new reduced financial allocations for each sector chosen in the first stage of the exercise were in the nature of ad hoc allocations to start the process of revising the Plan. Simultaneously with this part of the exercise, each of the sectors and ministries had set to work to re-order priorities in the light of the cuts that would have to be imposed.

It was recognised that a substantial curtailment of the investment programme would make it impossible to raise per capita income above the level of the benchmark year. Continuing low levels of income would continue to frustrate attempts to raise the savings rate substantially in the next plan period. Hopefully by the end of the decade, population planning efforts would start yielding results. This gloomy prognosis could be changed only if new resources like additional natural gas or oil were discovered or if a very substantial and continual flow of external resources from the rest of the world supported the economy. Viewed in this light the importance of investing in an intensive exploration of natural resources of Bangladesh was pushed to the fore as was the need to finance research and development efforts directed to the exploitation of known domestic resources. In the face of diminished resource availability, the importance of utilising existing capacity more fully rather than creating new capacity became increasingly apparent. Agriculture was to continue to receive high priority and there was the need to open up opportunities of small-scale private investment in the rural sector, both agriculture and related activities such as marketing of agricultural products, warehousing and storing, as well as transportation and trading, etc.

Within the agricultural sector itself, there was the need to postpone the slow-yielding, large-scale water control projects with long gestation periods. Greater emphasis had to be placed on labour-intensive, small-scale drainage schemes. The control of the major rivers through river training or multi-purpose embankments and barrages had also to be excluded from present considerations; instead drainage and dredging of small rivers and canals had to get priority. The excavation of derelict tanks which could help to provide a natural reservoir of flood water as well as provide supplemental sources of irrigation deserved a larger allocation of funds. This project was labour-intensive and required little or no foreign exchange. The coastal embankments which could prevent the incursion of tidal waves and saline water that obstructed the downstream flow of water from the rivers in the interior regions, were also to be given new emphasis together with maintenance, repair and improvement of those already constructed. Among the alternative irrigation projects, those which presented the least managerial and organisational problems were to be favoured. Large-scale gravity flow (canal) irrigation projects as well as tubewell development had to be slowed down in favour of low-lift pumps and shallow tubewells, which were not only less expensive but presented fewer organisational problems. At the same time greater attention had to be paid to rain-fed as against irrigated agriculture in terms of the supply of modern imports, including fertilizer, credit and extension service.

High priority for agriculture had the corollary that projects which were interrelated with it had also received emphasis. The case for expanding the steel mill which was unable to operate at its rated capacity was very much weakened. The project for a petro-chemical complex envisaged in the Plan had already undergone modification and only the fertilizer component of it had been preserved, with the possibility of adding to capacity in the future for export sales. Energy-intensive industry, given the high price of fuel, was accorded low priority, except for what promised to be profitable exploitation of the sulphur-free, good quality natural gas available in Bangladesh. It was also clear that the cotton textile spinning industry, which figured prominently in the Plan, needed to be reconsidered.

Among new projects in the industrial sector, that for producing cement was essential for development in other sectors of the economy. Uncertain and irregular availability of imports of cement, in the context of a limited construction season which required a well-programmed assured supply, combined with a rising trend in world price, added greater force to the argument for the exploitation of the domestic limestone even though it was relatively costly, provided technological problems of deep-mining could be solved. The alternatives of increasing domestic production of cement based on limestone imported from India, beyond the limits set in the Plan (already there was a plant operating on this basis) also assumed renewed importance. Plans for the expansion of the rest of the large-scale industrial sector were also casualties of the reappraisal. Rather than expand capacity, the hard core of priority projects would concentrate on selected modernisation, upgrading and balancing existing equipment and machinery with a view to securing optimum output from past investment. A much larger investment was to be made for repair and maintenance of existing and new equipment. This required investment in training and in repair and workshop facilities—a type of investment which had traditionally suffered from lack of funds. Public investment, in providing the physical infrastructure such as transportation and marketing facilities, as well as financial and technical assistance including a very elastic supply of credit and inputs, was a *sine qua non* of rapid growth of small-scale industries. Industries using domestic raw materials, for example, in leather goods, food processing. and a wide variety of engineering goods could be split up and shifted to the small-scale industrial sector.

A lower rate of agricultural and industrial growth required a lower rate of investment in the transport and communications sector. More than any other sector, this one was severely handicapped, faced with a backlog of accumulated requirements for replacement of obsolete and depreciated equipment and machinery, and lack of repair and maintenance facilities. Moreover, there was a need for training and development of manpower and professional skill. The results of the transport survey which were now available gave a new opportunity for a closer look at the priorities in this sector. Similarly, with a lower projected demand for power consequent on a lower rate of growth of the economy especially in the industrial sector, the focus of investment in this sector was also switched to repair and maintenance of the power facilities and supplementary investment for the replacement and balancing of equipment and machinery as well as training of manpower, in order to prevent frequent power failures and to utilise existing capacity fully.

In the social sectors the suggestion of the original Plan for drastically reducing public direct investment in housing, particularly in the urban areas, and replacing it by substantial expansion of the co-operative housing scheme, acquired new importance as did research on low-cost housing. In the field of education there was the need for a change in public opinion and in the attitudes of the leadership towards a desirable pattern of educational development in the future. In spite of many expert committees and commissions in the past, no clear guidelines seemed to be available. Nowhere else were political and ideological issues more explicitly intertwined.

Expansion of educational facilities along conventional lines, especially at the level of universities and colleges, was a clear case for curtailment, not only because of the scarcity of resources but also because it might create obstacles in the way of innovations because of the power of vested interests in higher education. This was not true of a considerable expansion of technical and vocational education which was closely integrated with the in-service training of workers in the industries. Moreover, a large supply of technically and vocationally trained intermediate skills might provide the hard core of entrepreneurs for small-scale industries, including repair and maintenance workshops. There was in fact a considerable scope for innovation and experimentation in the field of education. The Plan explored some but not all the possibilities. For example, if the political pressure for the expansion of primary education was to be met in the face of an acute scarcity of public sector resources, the local government could be entrusted with the responsibility of financing primary education with the help of local resources. To make the content of primary education biased towards agricultural subjects, to make urban students work in rural development projects during holidays and to integrate work on the factory floor with technical and engineering education, were a few examples of making education more work-oriented and directly linking it with development projects.

The Plan laid considerable emphasis on population planning. Slowing down the rate of growth of population was a vital objective. In the short run, any significant achievement in the latter was unlikely; what was critical was to establish institutions and policies which were likely to make a population planning programme effective within the next five years. In a revised smaller Plan no reduction in expenditures on this sector was desirable. The only valid limitation on the expenditures on population planning was the rate at which projects could be formulated and implemented. The most important aspect of the population planning programme during the next few years was the commitment of high-quality administrative personnel to this activity, who would establish and manage appropriate institutions and ensure the training of personnel. Experimenting with alternative institutional arrangements as well as with alternative techniques was very desirable. Even though a particular approach in terms of an integrated health and family planning programme was adopted, a readiness to adapt it as experience accumulated was needed. All available channels—private, voluntary, commercial or rural health centres—for the supplies of contraceptive devices should be experimented with. In order to ensure that there was no unsatisfied demand for family planning services, no opportunities should be left unexplored. In view of the lack of resources there would have to be a concentration on a limited number of health centres, each completed and run in a fully-fledged way as soon as possible so experience could be gained and experimentation with innovations appraised. In the health sector, the Plan had already emphasised training of para-medical personnel, medical assistants and nurses. In view of limited resources this would continue to form the bulk of the programme.

Notes

1. *The Five Year Plan*, p. 15. The "benchmark" output was different from the actual output in 1969/70 to the extent that it included extra output originating from the additional capacity which was created between 1969/70 and 1972/73. The rate of capacity utilisation was assumed to be the same as in 1969/70.

2. Ministry of Finance, *Annual Budget for 1971/72, 1972/73 and 1973/74. Annual Plans for 1972/73 and 1973/74 and Budget Speeches of the Finance Ministries*, each year.

From 1974/75 onward reconstruction expenditures were not separately shown in the budgetry accounts. In each sub-sector of programmes or projects, the agencies, however, did distinguish between the reconstruction component of their expenditures, which remained incomplete after 1973/74.

3. To make estimates for the years starting 1975/76 comparable with those earlier years, adjustments were made for the change in exchange rates which took place in 1975.

4. The Annual Plan of 1973/74, the first year of the Five-Year Plan, was finalised at the same time as the Five-Year Plan itself. That is why there was no divergence between the size of Annual Plan of 1973/74 and the phasing of the development expenditures as provided in the Five-Year Plan for the year 1973/74.

5. During 1973/74 the minimum level of non-development expenditures was estimated at about 40% higher than what was incurred during 1972/73. The increase was, however, mainly due to (*a*) the increases in expenditure on defence and law-enforcing agencies, which was very inadequate at the time of independence, and (*b*) increase in wages and salaries in the public sector.

6. Ministry of Finance, *Bangladesh Economic Survey*, 1975/76, p. 183.

7. *Five-Year Plan and Annual Plans*, op. cit.

8. *Annual Plan*, June 1973, p. 1–2.

9. The per capita consumption was about 322 lb during the war of liberation in 1971/72, but then for more than half of this year, 10 million out of 72.4 million population were refugees in India which provided for their food consumption. If about 10–14% of the total consumption requirement was assumed to have been supplied by India for half the year in 1971/72, then per capita consumption during 1971/72 would have been between 350 lb and 357 lb.

10. *Annual Review*, 1974/75, pp. 30–36.

11. *Annual Review*, ibid.

12. *Annual Plan*, 1973/74, pp. 59–60, and 1974/75, pp. 30, 204.

13. *Bangladesh Economic Survey*, 1975/76, p. 114.

14. The impact of changes in income and its distribution were considered unimportant during the short period under review.

15. *Annual Plans*, 1972–75, ibid.

16. The stocks of raw jute and jute goods at the end of the year are shown in the table for each of the years:

	Year	*Raw Jute (000 bales)*	*Jute Goods (000 bales)*
30 June	1970	530	63
	1973	684	101
	1974	691	114
	1975	654	127

Source: Bangladesh Jute Industry Corporation, unpublished data.

17. *Annual Plans*. 1972/73 to 1976/77.

18. Government of Bangladesh, *Economic Review*, 1974/75 and *Economic Survey*, 1975/76; *Annual Plans*, 1974/75 and 1975/76; *Memorandum for Bangladesh Aid Group*, 1975/76 and 1976/77.

19. The fall in terms of trade over the years is shown as follows:

PRICE INDICES AND TERMS OF TRADE

	Export Price	*Import Price*	*Terms of Trade*
1972/73	100	100	100
1973/74	104.6	156.9	66.7
1974/75	132.9	215.4	61.7
1975/76	111.6	193.5	52.7

Source: Government of Bangladesh, Planning Commission, *Memorandum for Bangladesh Aid Group*, 1975/76 and 1976/77.

20. *Annual Plans*, op. cit.

21. *Annual Plans* and *Memoranda for Aid Group*, various years.

22. *Bangladesh Bank Bulletin*, 1973–75 (various issues).

23. *Annual Plans*, op. cit., Bangladesh Bank and Planning Commission, unpublished data.

24. ibid. Between 1969/70 and 1973/74 domestic prices increased two and a half times more than comparable price indices in the major trading partners. Against this, the Bangladesh rate of exchange was devalued by only some 58%.

25. Bangladesh Bank.

26. Figures for 1972–75 are the averages for the two months in each year. World Bank, *Bangladesh Current Economic Performance and Short-Term Prospects*, 22 March 1976.

27. *Annual Plans* and *Bangladesh Bank*, op. cit.

28. *Report of the Overseas Development Ministry of the U.K. Mission on Adamjee Mills*, 1974.

29. *Report of the Japan Textile Machinery Engineering Mission*

submitted to Bangladesh Textile Industries Corporation. Also the *Report of the Bangladesh Textile Industries Corporation,* April 1973, unpublished memorandum.

30. *Annual Plan,* 1973/74, p. 53.
31. *The Annual Plan,* 1974/75, pp. 150–2.
32. *Annual Plan,* 1974/75, pp. 118–19 and 131.
33. ibid., p. 119.
34. *Annual Plan,* 1974/75, pp. 76–81.
35. IBRD, *Bangladesh: Development in a Rural Economy,* Vol. 1, September 1974 (unpublished), p. 298.
36. Faaland and Parkinson, op. cit.

6. Domestic Resource Mobilisation

(a) Introduction

How to accelerate the rate of mobilisation of domestic resources for financing development was one of the key questions in economic policy-making and planning efforts in Bangladesh. The problems and prospects of domestic resource mobilisation involved a wide range of complex and interrelated political and economic considerations. This included not only problems of choice between present and future consumption but also of distribution of income and allocation of resources between the different sectors of the economy, between the rural and urban areas, and between different income or occupational groups in the country, all of which had differential political strength or exercised different degrees of pressure or influence on economic policy-making. The earlier chapters on the planning process have explained the interrelationship between planning, on the one hand, and institutional, administrative and political factors, on the other. The following discussion of the evolution of Bangladesh's policies towards the mobilisation of domestic resources focuses mainly on the interplay of economics and politics, i.e. interaction between various interest groups in society. This was an area of policy-making in which political limitations on economic policy-making and their complex interrelationships became very evident over time; this chapter seeks to explain how Bangladesh attempted to deal with this complex problem, which was critical for the future development of the country.

The policy questions which centred around the problem of domestic resource mobilisation were at least threefold: firstly, its relationship to foreign economic assistance; secondly, its relation to the size of the development programme; and thirdly, the question of the relative shares of public and private saving.

There was both a complementary and a competitive relationship between foreign economic assistance and domestic resource mobilisation.[1] Starting from a very low level of domestic resource mobilisation, an increase in domestic saving was essential for utilising or absorbing an increasing amount of foreign economic assistance for development purposes; food and other assistance to support consumption levels needed no domestic saving effort. The utilisation of project assistance required local expenditure on construction and building, and on wages and salaries of labour and other staff engaged on the implementation of the project. Unless the counterpart expenditures could be financed by local resources, the available project assistance might remain unutilised or under-utilised. However, this was mitigated in so far as some foreign assistance was available to finance also the local costs of foreign aided projects. Foreign assistance and domestic resource mobilisation were complementary in yet another sense. Frequently, self-help in the sense of an acceleration in the rate of domestic saving, was considered by donor countries as an important performance criterion in the allocation of foreign economic assistance: a recipient country which did its best in mobilising domestic saving deserved greater foreign assistance. Foreign aid, in other words, was used as leverage in stimulating greater domestic savings efforts, and also as a guarantee that a receiving country would not become a long-term or permanent liability on the generosity of the donors. The Planning Commission and the Finance Ministry frequently used this agreement in negotiations with the Cabinet for new and additional tax measures.

The fact that domestic resource mobilisation and foreign economic assistance were substitutes, as well as complementary to each other, created a dilemma for policy-makers. Increased foreign assistance was not always or necessarily an easy way out of burdensome domestic efforts and austerity: larger domestic efforts helped bring in larger assistance and enabled its productive utilisation. Moreover, while increased taxation and imposition of sacrifice on domestic consumption was clearly politically unpopular, heavy dependence on foreign assistance was also politically unpalatable because of the constraints on freedom of action imposed by heavy dependence on foreign aid. Such dependence on foreign aid not only could restrict freedom of action or autonomy in external and internal policies, it might also act over time as a self-defeating policy in so far as slow progress towards self-reliance would eventually discourage donors and turn off the flow of aid.

The second aspect of the strategy of resource mobilization was that the lower the rate of domestic saving, other things remaining constant, the lower was the size of the development programme. The policy-makers faced a difficult choice in this respect. While increased taxation and public savings imposed sacrifices on current consumption, lower levels of development expenditures, caused by inadequate savings, restricted the growth of income and employment and, therefore, imposed sacrifices for the future. There was great pressure for increasing development expenditures, especially in view of rising population, on the one hand, and the very low level of social services, on the other. It was hoped that resources mobilised through the curtailment of non-essential private consumption would not be taken up by non-productive public expenditure on general administration and defence services, etc., but would be invested in socially beneficial and productive projects, which would contribute to growth and equity.

The problem of choice for the policy-makers was particularly difficult because while, on the one hand, effects of increased taxation were felt immediately by the masses, the impact on the welfare of the masses of increased development expenditures, financed by increased taxes, could only be felt after a time lag; from this point of view there was a strong case for increasing the direct link between increased taxation, on the one hand, and increased development expenditures, with relatively short non-impact on the common conditions of the masses, on the other. The benefits of public expenditures should be readily perceived by those who bore the burden. For example, mobilisation of resources for the public sector at the level and for the use of local governments for clearly identifiable projects which were of immediate and direct benefit to the local community had a greater appeal and political acceptability. However, the extent of devolution of financial resources and responsibility for development projects to the local authorities was very limited in Bangladesh. The local government institutions had suffered from a serious setback during the war of liberation.

The third aspect of resource mobilisation was concerned with the relative shares of public and private saving. This aspect of the problem was particularly relevant in the context of Bangladesh because of the overwhelming share of the public sector in total development expenditures and the consequent need to mobilise private savings for investment in the public sector. Of public sector expenditures in the Five-Year Plan 12% were to be financed by borrowing from the private sector.[2] Almost half of the total private savings were to be mobilised for investment in the public sector. This implied that a significant proportion of private savings was to be held in the form of financial assets provided by the financial intermediaries and the government.

The relevant policy issues in this context related to the optimum choice of instruments and incentives for mobilising private savings for use in the public sector. This involved the role of the financial institutions and of the credit and interest rate policies for the mobilisation of private savings in the form of financial assets. During 1973/74 and 1974/75 about 4% to 5% of the public sector development expenditures was financed by borrowing from the private sector by means of various types of savings certificates, prize bonds, postal savings, postal life insurance and provident fund contributions. It was not certain that the level of domestic savings was independent of the preferences of the savers as between the alternative assets in which savings were to be invested. Past experience frequently suggested that private savings, especially in the rural sector and among the surplus farmers, were often linked up with opportunities of direct investment. Ready availability of small-scale, highly profitable investment opportunities stimulated private savings, especially among the small savers. This raised the question whether the allocation of development expenditures between the private and public sectors, as postulated in the Five-Year Plan, was consistent with the relative sources of saving. In other words, the achievement of the target of domestic savings might require a greater scope for private investment, if private savings were not to be discouraged, or if private capital was not encouraged to migrate abroad. This issue became a matter of considerable concern as the implementation of the Five-Year Plan proceeded.[3]

Three important potential sources of capital accumulation in Bangladesh were agricultural surplus, industrial profits, and trading profits (including profits made on imports as well as from domestic marketing and distribution of agricultural and industrial commodities). The agricultural sector, the largest sector of the economy, was a potential source of capital accumulation, particularly from the operation of the surplus farmers. Those agriculturists who adopted modern technology earned high profits, aided by subsidies on inputs such as fertilizer, pesticides, seeds and water for irrigation. The subsidies were a potential source of savings to the extent that subsidies were no longer necessary in order to stimulate the adoption of new technology in agriculture.

Since an overwhelming proportion of the industrial sector was nationalised, the profits generated by the nationalised sector constituted the main source of industrial profits. Profits in the foreign trade sector were shared between the public trading agencies and the private importers. However, even in the case of public sector imports, private intermediaries, popularly known as indentors, acted as agents of the suppliers and, therefore, shared in the profits of the import trade. There were many importers and trading intermediaries and they were widely dispersed. There was a significant decline in the degree of concentration in the import trade in the post-independence period because the large trading houses in the pre-independence period were mainly of Pakistani origin, and to replace large abandoned Pakistani trading houses, there was a deliberate attempt to distribute import licences amongst a large number of importers. The small-scale operations did result in high costs. The distribution, both wholesale and retail, of both imported and domestically-produced commodities was predominantly in private hands, except for a few rationed commodities and a shortlived experiment with state trading in the form of "fair price" shops. Throughout the period under review there was an acute shortage of both imports and domestically-produced manufactured goods. There were high scarcity profits per unit of traded commodity and they were unevenly distributed amongst the trading intermediaries. Because of imperfection in the capital market, and lack of knowledge and expertise in the import trade, the original recipients of import licences often sold them to a smaller number of importers, who had access to capital and the knowledge to engage in the actual task of import trade. Even though there was a great scarcity of domestically produced commodities, especially manufactured goods, the pricing policy of the mainly publicly owned industrial enterprises was not designed to mop up the scarcity margins. They instead accrued to the trading intermediaries.

In terms of either average per capita income or consumption of items of basic needs, Bangladesh was a country of great poverty. But in a general overall picture of poverty and deprivation, there were pockets or sectors of the economy generating surpluses which could be mobilised for investment. The problem was to devise and implement the most efficient and equitable methods of mobilising such surpluses and to ensure their most efficient use so that Bangladesh could increase the degree of self-reliance in the foreseeable future.

With a large and growing population, she had a large surplus labour force available for mobilisation for productive work. How to mobilise this surplus labour without a corresponding increase in the supply of wage goods, on the one hand, and capital equipment, on the other hand, was the biggest challenge facing the planners in Bangladesh. Bangladesh had neither the benefit of a well-organised cadre of workers nor an ideologically oriented authoritarian political system to meet such a challenge. Nonetheless to energise and motivate the impoverished masses for productive work with nothing more than a promise of very slow growth in levels of living was something which Bangladesh had to achieve if she was to make progress.

(b) Techniques of Resource Mobilisation

Among the instruments which Bangladesh had at her disposal for mobilising investment resources, the most important were taxation, reduction or elimination of subsidies, and the generation of surpluses by the nationalised enterprises. The problems of generating surplus in the nationalised enterprises which were linked up with the optimum methods of their organisation and management are discussed in the next chapter. Deficit financing, i.e. creation of bank credit to finance government expenditures, was always an alternative to taxation. Because of heavy reliance on indirect taxation the incidence of which fell principally on the items of mass consumption, the regressive effect of deficit financing was not very different from that of increased indirect taxation.

Faced with a choice between the creation of bank credit and increased taxation for financing public expenditures, there was a tendency to use deficit financing in preference to taxation. The incidence of taxation in terms of those who had to bear the immediate burden was direct and identifiable whereas the incidence of deficit financing was widely diffused. The impact of inflation on the volatile and organised pressure groups like the industrial workers could partly be mitigated by higher wages. Moreover, resort to the creation of bank credit as a means of financing public expenditure enabled the government to finance a few items of expenditure without having to include them in the budgetary accounts; this under-estimated total deficit in the fiscal operations of the government and it consequently underestimated the impact of the aggregate fiscal operations on the supply of credit and the inflationary potential in the economy. A good example was the way in which the food subsidies and the deficits of the nationalised trading and industrial enterprises were handled in the accounts of the government. They were represented as a part of the financial transactions of the banking system with its clients rather than as consequences or components of the governments' own fiscal operations and deficits. The accounts relating to the deficits in the current or revenue budget of the government did not include borrowing by the government from the banking system to provide food subsidies or to meet the deficits of the governments' trading or industrial enterprises.

(i) *Taxation.* Two principal areas of taxation in Bangladesh, as already pointed out, were profits earned in the marketing and distribution of both imported and domestically-produced goods and income, or surplus generated within the agricultural sector.

The prevailing rates of import duties or excise and sales taxes left room for additional taxation of profits from the distributive trade. The average duty in Bangladesh was about 30% of the value of dutiable imports; the luxury and non-essential goods, however, were taxed at higher rates. For example, cigars, cigarettes and cosmetics etc., were taxed at 100–150%, household appliances at 100–150%, fine cotton fabrics at 125–200%, etc. Essential goods like foodgrains, fertilizers, seeds of all types, metallic ores and infant foods were free of duty, but duty-free goods on the whole were very limited. There was the intermediate range of essential goods such as coarse fabrics, pharmaceuticals and industrial raw materials which had relatively low rates. Machinery and machine parts also fell in the intermediate range. In view of stringent foreign exchange control, the relative proportion of non-essential imports in Bangladesh was very small. Accordingly, items with high import duties, which were non-essential items, did not contribute much to the total revenue derived from import duties. With low rates or no duty on essential imports, any attempt to increase substantially the yield from import duties would have required an increase in the rate of duties on a wide variety of raw materials and capital goods. Similar was the case with excise and sales taxes. The average rates of these taxes constituted about 10% of the combined gross value of output of the large-scale industry. If the output of the small-scale and cottage industries, most of which were exempted from excise and sales taxes, were considered, then the ratio of these indirect taxes to the total industrial output would be much smaller.

There was, however, an important argument for increasing the import duties across the board or for "auctioning" of import licences. This was due to the fact that the domestic prices of imported goods exceeded by a wide margin the duty-paid value of imports, in view of an excess demand for imports suppressed by stringent import and exchange control. The advantage of low duties was not secured by the ultimate users or by the intended beneficiaries of the system of low import duties. The scarcity margins ranged between 100% and 150% on the duty-paid import price or the tax paid ex-factory price in the case of a wide range of commodities. A few examples as of mid-1974 are quoted opposite.

At an early stage in 1973, the proposal of auctioning import licences was considered as a method of siphoning off the excess profits on import goods. In the case of imports which were directly imported by the state trading agency, the auctioning of imports in the open market was suggested as the appropriate procedure. Various objections were raised against the auction of import licences. First, it was feared that auctioning would enable a few importers with large financial resources to obtain the bulk of the import licences. This would result in a concentration of the import trade in a few hands, which was undesirable from the point of view of income distribution. However, this could have been counteracted by putting a limit on the amount of licences which any single importer could bid for. Secondly, a system of auctioning could upset the equitable distribution of opportunities of engaging in import trade among the different regions of the country, which was built in the import licensing system, unless this was counteracted by earmarking or separating auctions for different regions. The limitations such as those above on the working of the auctioning system would have interfered with the attempt to siphon off excess profits to the fullest possible extent. But still it would have gone a

SCARCITY MARGINS ON SELECTED COMMODITIES
PERCENTAGE MARK-UP ON EX-FACTORY OR IMPORT PRICE

	%
Sugar	100–150
Cement	300–400
Trucks	100
Milk Food	150
Razor Blades	400
Condensed Milk	300–400
Tyres and Tubes	100
Coconut Oil	200
Cigarettes	80–100
Television Sets	200–300

Source: Planning Commission; also *Report of the Working Group of the Officials' Committee on Distribution of Essential Commodities*, unpublished memorandum, 1974.

long way in mobilising resources. The most important argument against the proposal was, however, that it would not only mop up scarcity margins but also would raise the market price by raising the cost of imports to the importers. It was contended that the importers, wholesalers and traders did not charge the highest prices that the market could bear and that they fixed market prices on the basis of cost plus a fixed margin. Therefore, any rise in cost, given the percentage margin, would raise prices. Under the system of pricing based on cost plus fixed margin, there was always an excess demand at the existing ruling price, which went unsatisfied. The incidence of the unsatisfied demand on the consumers was random. Those among the consumers who were lucky to obtain access to supplies enjoyed the additional consumers' surplus.

There was a related argument against auctioning import licences. The government had sought to fix the prices of a large number of both domestically-produced and imported commodities. Even though the system of price control was not enforced by a system of control over supply and distribution, for at least a few commodities always, and in the case of many commodities occasionally, the consumers had access to supplies at or near controlled prices. The assumption was that not all sellers and certainly not all of them maximised profits all the time and consistently took advantage of domestic scarcities. If the cost of imports went up as a result of auctioning import licences, consumers' access to supplies at controlled prices, however limited and infrequent, would totally disappear. This line of reasoning was a mixture of political and economic considerations, perhaps mostly political. The government was anxious to demonstrate that it was very concerned about the high prices paid by the consumers, and that its attempt at fixing prices of essential commodities was evidence of its concern. The fact that the ruling prices were above the controlled price could be declared to be due to the manipulation of the market on the part of the traders who were blamed for exploiting the consumers. It was argued that if there was excess demand, the traders should not attempt to charge scarcity prices but should institute a system of rationing on the basis of "first come, first served", rather than allowing market price to curtail demand to the level of available supply. The government appeared to believe that some traders, if not all, behaved in this fashion, and that there was no reason why all traders should not behave likewise.

The same line of reasoning was advanced against an increase in import duties as well as in excise and sales taxes on imported goods. The hypothesis that an increase in taxes and duties would not lead to a rise in prices under conditions of domestic scarcity was not acceptable. In any case, there was no way of testing the hypothesis from historical data, because simultaneously so many other changes in conditions of supply and demand took place.

The kind of considerations which entered into the determination of taxation policy were amply demonstrated by the experience of the Annual Plans of 1973/74 and 1974/75. During these years, a number of new taxation measures were under consideration. The most important proposals related to an increase in import duties on cotton and cotton yarn, and an increase in excise duties on petroleum products. The additional taxation on cotton yarn and raw cotton was opposed on the grounds that the incidence of increased taxes would be on the poorer sections of the people, even though there was a wide margin between the import cost and the prevailing market price. Moreover, the percentage increase in the cost of the finished textiles would have been only a fraction of the percentage increase in the cost of raw cotton and cotton yarn. The strongest opposition to increased taxation on cotton yarn was based on the grounds that cotton yarn was used predominantly in the handloom industry which was the most important source of non-agricultural income and employment for the rural poor. That the price paid by the handloom weavers was much higher than the ex-factory or import price of cotton yarn and that the traders reaped the scarcity margins on cotton yarn, were not denied. To ensure a distribution of cotton yarn to the weavers, at cost plus reasonable profit, an attempt was made to organise weavers' co-operatives which were to directly receive quotas of cotton yarn from the government agency, Bangladesh Small Industries Corporation, or from the large-scale, nationalised cotton textile industry, in proportion to the numbers of weavers included in the co-operatives. However, weavers' co-operatives in many instances turned out to include a large number of traders or marketing intermediaries. In other words, co-operatives turned out to be societies of middlemen-cum-traders who obtained cotton yarn at prices fixed by the government and sold it on to the weavers at scarcity prices, i.e. with high profit. The administrative problem of ensuring that only genuine weavers' co-operatives secured access to cotton yarn at controlled prices proved insurmountable. The weavers remained greatly dependent on the middlemen–traders for the supply of credit and for the marketing of their products. Similarly an attempt to distribute cotton yarn directly to the individual weavers under a system of rationing by the state trading agencies, including state-owned retail shops, also failed. The task of obtaining a reasonably accurate census of weavers and supplying them with ration cards was soon found to be administratively not feasible.

The proposal for enhanced taxation of petroleum products was accepted, with the important exceptions of kerosene oil and high-speed diesel oil, for which there was government control over supply and distribution. Kerosene oil was the common man's fuel for domestic use, whereas high-speed diesel oil was the fuel for running irrigation pumps which were considered essential for the agricultural development programme. Except when the supply of kerosene was extremely short, the urban consumer secured a large part of his requirements through the ration shops. Control over the distribution of diesel oil for use by the irrigation pumps was, however, less satisfactory.

Thus the proposal for enhanced taxation of raw cotton, cotton yarn, kerosene oil and diesel oil was not acceptable. This was either because it was believed that in spite of scarcity margins earned by the traders, an increase in taxation would increase prices to the final consumers and would have adverse income distributional effects or would provide disincentives to production. To some extent it was believed that profits earned from trade in these commodities were not very high, especially in view of the wide distribution of import licences. The moderate profits earned by a very large number of recipients of import licences distributed income to the low- or middle-income traders, many of whom had suffered losses of income and assets during the war of liberation; there were also many among the traders who provided considerable support and strength to the ruling party. The resistance to increased indirect taxation on imports or domestic outputs was not without exceptions. To illustrate, import duty on cement was raised in 1973/74. Partly this was a price-equalising device; domestic output was more expensive than imported cement. A further consideration was that the burden fell on the middle-income groups because it was they who undertook the bulk of the urban residential construction. The crucial reason, however, was that this was one commodity for which the market price was several times higher than the ex-factory price or import price and that the government showed some readiness to tax scarcity margins only when the scarcity margins were exceedingly high.

The acute scarcity of financial resources as well as the need to restrain excessive deficit financing required the government from time to time to consider seriously proposals for taxing excess profits on the traded commodities. Even though proposals for an enhanced excise duty on cotton yarn and cotton cloth were rejected in 1973/74, they were again mooted in 1974/75. At that time, a new proposal was put forward for a 20% import licence fee for all imports. This was accepted but with a number of exemptions. The exemptions included such commodities as foodgrains, fertilizers, crude and refined petroleum, insecticides, edible oil, baby food, raw cotton, cotton yarn, cotton textiles, refined coconut oil, medicines and non-developmental imports on government account. The list of exempted commodities was chosen with a view to protecting the prices of what were considered essential commodities. At the same time, the state importing agency, Bangladesh Trading Corporation, had to pay for the first time a turnover tax of 5% of its total annual business turnover. The declared intention was to mop up a part of the scarcity value of imports in the form of revenue and to make the corporation more cost conscious.[4]

The Cabinet, whenever considering additional measures of indirect taxation, whether of sales tax, import duties or excise duties, always insisted on being given a precise figure for the quantitative impact on prices of such measures. It was one thing to work out the impact on the ex-factory cost or on the landed cost of imports of additional import duties or licence fees and sales tax, and quite another to quantify the impact on the ruling market price in which there was an element of scarcity premium. Moreover, the impact of the increase in import duties or excise/sales taxes on the price ruling in the market could not be predicted because of changes in market situation affecting the supply and demand of the taxed commodities. For example, the price or the supply of imports did not always remain constant after the imposition of an increased import licence fee or import duty; other things remaining equal, a smaller supply owing to reduced imports would necessarily cause a rise in prices. Similarly, an upward shift in demand or an increase in traders' stocks or consumers hoards would put upward pressure on prices. Furthermore, in response to a rise in import duty or surcharge, the importer or trader could push up the price in the short run to the full extent of the duty, even though this initial increase in price could not be sustained when its effects on demand had had time to work themselves out. However, the short-run increase in prices did not always get corrected after a time perhaps because scarcity margins had not previously been exploited or because other factors affecting the supply and demand intervened. In any event, there was always the possibility of a speculative element which generated the expectation of a future scarcity and caused a withdrawal of supply from the market in the hope of more profitable sales later on. An analysis of past experience relating to the movements in prices and in taxes could not be expected to disentangle the net influence of all these and others acting on prices; therefore no conclusive evidence could be advanced that the existence of scarcity premia on traded commodities guaranteed that proposed additional taxes would not raise market prices. In a general inflationary situation an increase in taxes was likely to strengthen the impact of other price-raising forces. It was of no practical consequence for policy-makers to be told that if scarcity margins prevailed, and if in fact the traders already charged the highest prices that the market could bear, and if other things did not change, the additional taxes would not raise prices.

The dilemma with which the government was constantly faced was the conflict between the revenue-yielding potential of taxes, on the one hand, and their income distributional effects, on the other. Taxes which yielded large revenue were also those which were imposed on commodities of common use or consumption by the masses and hence their incidence fell disproportionately on the poor. It was in this context that a number of new measures were mooted in 1973/74, which were not expected to yield much revenue, but of which the incidence would have been on the rich and affluent. It was argued that these measures, by requiring the higher income groups to accept a cut in consumption at a progressive rate, would impart an increased element of equity to a tax system, to which the masses of the poor made the largest contribution. These tax measures were mainly designed to tax high consumption expenditures or unearned income and windfall profits. They were the following: (*a*) tax on hotel bills in luxury hotels in cities; (*b*) tax on consumption of electricity by the domestic consumer; (*c*) tax on water consumption in the cities; (*d*) tax on first-class fares on railways and inland water transportation; (*e*) increased sales tax on jewellery; (*f*) tax on house rents; (*g*) increased registration fees for contractors; (*h*) increased registration fees for importers and indentors; (*i*) increased registration fees for lawyers and medical practitioners.

The fees for importers and marketing intermediaries as well as for medical and legal practitioners were intended to mop up the excess profits earned by them, especially since with inadequate information and enforcement machinery the income tax system did not succeed in bringing a large part of their income within its net. Though original proposals as mooted by the planners were intended to introduce new taxes at high rates or to substantially increase the existing ones, a few of their proposals such as tax on consumption of electricity and water, travel by railways and waterways, purchase of jewellery and house rents were rejected, and in the case of a few others only modest increases were introduced. The proposal for licence fees on the marketing intermediaries, contractors and importers was accepted but at a much lower rate than was originally suggested on the basis of the same reasoning which was applied against either enhancing import duties or auctioning import licences. As far as the medical and legal practitioners were concerned, the original suggestion was that of a considerably enhanced registration fee combined with a surcharge which would be related to the number of years of practice, with a higher surcharge for the urban areas than for the rural areas. This proposal was rejected on the grounds that in the post-independence period the real income of these professional groups had already declined due to the high rate of inflation and that any new tax measure would not only discourage the growth of a professional class like doctors, of which Bangladesh was in short supply, but would also encourage emigration, which was already substantial amongst the medical practitioners. However, there was indeed a dilemma in this respect because there were no doubt many medical and legal practitioners who earned high incomes, especially in the metropolitan cities, and who avoided direct income taxation. The case of legal practitioners was a little more complicated. There was considerable overcrowding in this profession and there were many in the small urban and rural areas who did not earn income on their past investment in education in any way comparable to the income from other professions involving similar investment. To make matters complicated, most of the members of the ruling party and the party officials were drawn from the legal profession. A special tax on the legal profession, while it might have had the desirable effect of discouraging new entrants, was likely to hit adversely the large number of legal practitioners, many of whom were not affluent and many of whom supplied the hard core of leadership of the political parties; it would have been a very unpopular act.

It may be instructive to illustrate the range of considerations, varied and sometimes conflicting, which affected the taxation policy in Bangladesh. For example, the objective of augmenting government revenue by means of increased indirect taxes was not always consistent with the objective of securing the most efficient use of scarce foreign exchange or capital or the most socially desirable composition of imports and output.

To illustrate: in the interest of conserving scarce foreign exchange, Bangladesh had imposed a ban on the import of cigarettes and alcoholic beverages. The question was debated as to whether it was desirable to continue the ban which was partially violated through illegal imports or smuggling, which in turn resulted in a loss of both revenue and foreign exchange. Legalising imports would keep the net impact on the foreign exchange resources unchanged but would, on the other hand, add to the revenue resources of the government, unless imports were to be allowed in excess of what came in previously through illegal channels. Since illegal imports were not quantifiable, one could not be sure of this; however, legally licensed imports could be kept at a level which was much lower than what would occur if imports were to be completely liberalised. The problem of choice, in this case, was a political one. The government was anxious to create an impression of austerity in the country so that the high income groups were not seen to indulge in luxury consumption. The fact that the smuggled cigarettes were on sale in the market was a source of embarrassment to the government. Often attempts were made to forcibly seize or destroy such smuggled goods.

It was felt that the inability to stop the smuggled imports of luxury items was an admission of failure of the government's enforcement machinery and that drastic steps should be taken to eliminate smuggling. However, in practice it was found that it was impossible to enforce such a ban. After a year or two, it was decided to allow imports of a very limited amount of cigarettes under very high rates of import duties.

The case of excise duties on domestically manufactured cigarettes was even more complicated in terms of the conflicting considerations which were involved. The cigarette manufacturing industry was the largest source of revenue from excise duty, both because the tax base was large and the tax rate was high. However, it was a capital-intensive, highly mechanised industry and was partly foreign-owned and managed. Also, it depended on imports of high quality tobacco to be used in combination with domestic tobacco and on imported wrapping paper for cigarettes. It competed with the indigenous cottage industry, producing an inferior substitute called "bidi", entirely from domestically produced tobacco with the use of very little capital. The expansion of the indigenous cottage industry at the expense of the modern industry would have produced more employment, absorbed less capital and foreign exchange resources. It was true that a specialised variety of tobacco leaf which was used as wrapping material for the indigenous cigarettes was often smuggled from India, but a domestic substitute for this wrapping material was being increasingly produced in Bangladesh. While the expansion of the indigenous industry would have saved on capital and foreign exchange and provided much greater employment, it would have yielded no revenue. This was because of the nature of production in the cottage industry; it was mainly a household industry, employing both women and children, scattered throughout the rural areas of Bangladesh. It was difficult, if not impossible, to administer a system of excise duty on the indigenous cigarettes. Also, because of the very small scale on which trade in this commodity was carried out, no sales tax levied on it could have been effectively collected, nor was it administratively feasible to collect and levy a duty on the raw material, i.e. domestic tobacco produced on small farms all over the country. It was far easier to administer excise duty and sales tax on modern cigarettes, which were collected at the factory gate of a few large-scale enterprises. It was also easy to tax imported tobacco and wrapping paper used in the modern industry through the levy of import duties. Therefore, given the fact that Bangladesh had already inherited considerable installed capacity in the modern cigarette manufacturing industry, the easiest way of collecting large revenue was to allow liberal imports of raw materials and spare parts for the fullest utilisation of this capacity. In fact this was the policy which was followed even though it was an inefficient use of foreign exchange from the point of view of income and employment.

There were, however, other important cases where the objective of increasing foreign exchange earnings or conserving foreign exchange resources was in conflict with the objective of increasing government revenues. Since rice and jute competed for the use of land, an increase in jute production implied a corresponding fall in rice production. To the extent that an increase in foreign exchange earnings from the exports of jute and jute products more than compensated for the rise in foreign exchange expenditures on increased imports of foodgrains, there was an increase in dutiable non-food imports and consequently, in customs revenue. However, this process of substitution in favour of jute and against rice would be economical only so long as increased exports of jute and jute goods did not cause a fall in their export price to such an extent as to offset the favourable price ratio of jute *vis-à-vis* rice in the world market. This reduced the amount of net foreign exchange resources available for non-food imports and consequently reduced the tax receipts on such imports.

The exchange rate policy was closely related to the structure of duties on exports and imports and, therefore, to the total customs tax revenue. The maintenance of an over-valued exchange rate in Bangladesh provided an opportunity for the use of import duties to mop up scarcity margins and to generate large amounts of customs revenue. However, the devaluation of the exchange rate in 1975 reduced the scarcity margins and hence the scope of taxing excess profits on imports. But on the other hand, in the post-devaluation period, receipts from ad valorem import duties went up and the rates of import duties on a number of items were adjusted downwards in order to keep their prices from rising too fast or too high. At the same time, the devaluation reduced the need for an export subsidy and, in the case of the jute manufacturing industry, the need for production subsidies as well. With higher export receipts in local currency per unit of export of jute goods, it was possible to offset the high cost in the jute manufacturing industry.

As stated earlier, agriculture which was the largest single component of gross domestic product, provided an important potential source of surplus for financing development in Bangladesh. In the first instance, it should provide resources at least for accelerating the rate of investment within the agricultural sector itself. The agricultural sector carried a relatively low burden of direct taxation. In spite of a considerable rise in the price of output and in the profitability of agricultural operations in recent years, the rate of land revenue had remained unchanged. There was an agricultural income tax which was limited to income from agricultural land holdings above 33.3 acres (100 bighas) with a net income of Tk.4,000 and above per annum. After the imposition the ceiling of land ownership at 33.3 acres, the agricultural income tax had become irrelevant, except in the case of large agricultural estates such as tea estates.

Two changes in the system of land revenue and agricultural income tax were introduced by the government of Bangladesh. One was that the collection of agricultural income tax was postponed for two years from 1973/74 on the grounds that agricultural production suffered from dislocation during and immediately following the war of liberation. Considering the high level of income at which tax was imposed and the low rate of taxation, it was questionable whether such relief was necessary. The second step was to abolish land revenue on landholdings below 8.3 acres (25 bighas) in size. This was one of the electoral commitments of the ruling party made in the pre-independence days and was based on two grounds. Firstly, the system of collection of the land revenue by the local revenue officials involved considerable harrassment of the farmers as well as corruption on the part of the officials. Secondly, while the rate of land revenue was very small, i.e. about Tk.3 per acre, the cost of collection was a very large proportion of the total gross yield from land revenue. In the late 1960s the administrative expenditures of the land revenue department constituted about two-thirds of the total yield of land revenue. However, the administrative expenditures of the revenue department were not related entirely to the task of collection of land revenue. The department performed functions such as land surveys, maintenance and updating of land records, collection of fees and dues from fisheries, marketing places and bazaars. While land revenue was abolished in the post-independence period, there were a number of taxes or fees such as development and relief tax, education fees and local rates levied at rates which were expressed as percentages of land revenue. The latter served as "tax base" for the estimation and collection of these miscellaneous taxes. Even though these fees and rates were not abolished, their collection drastically declined in the wake of the abolition of land revenue. This was because the taxpayers defaulted on payment on the plea that the basis on which they were collected no longer existed. Exemption of landholdings below 8.3 acres in size exempted 90% of the agricultural land and 99% of the rural population from land revenue.

The abolition of land revenue on landholdings below 8.3 acres and the very low rate on large holdings coupled with an ineffective agricultural income tax, raised the basic question as to the relative burden of taxation on agriculture. A related question concerned the most efficient and equitable way of taxing agriculture. There were a large number of surplus farmers who had annual incomes higher than the minimum exemption limit for income taxpayers in the non-agricultural occupations. In 1973/74 the exemption limit for income tax payment was an annual income of Tk.6,000. Per acre net income from agriculture at that time varied between Tk.1,000 and Tk.3,000, so that a farmer with holdings of about 5 acres had a net income between Tk.5,000 to Tk.15,000 per year. A farmer with 8.3 acres and above, earned between Tk.8,000 and Tk.24,000 of net income per annum and paid at the prevailing rate of land revenue of Tk.3 per acre, only a total amount of Tk.25 which was between one-third and one-tenth of one per cent of his net annual income.[5] This was considerably less than the direct tax burden on non-agricultural income. It was true that the agricultural population, like their urban counterparts, did pay indirect taxes such as import duties, sales and excise taxes on what they purchased of both imported and domestically manufactured commodities. However, since the rural population consumed a relatively smaller proportion of such commodities, the incidence of the indirect taxation was likely to be lower on them than on the urban population. Against this, the overvalued exchange rate of Bangladesh until May 1975 when Bangladesh devalued its exchange rate, was a method of indirect taxation of agriculture, since it reduced the receipts from agricultural exports below what they would have obtained with an equilibrium exchange rate. While this was indeed the case for jute, the rice growers obtained prices which were higher in fact than the world market prices. Rice production like manufacturing was an import-substitution activity and hence received higher prices due to the overvalued exchange rate. The agricultural sector in the post-independence period not only paid very little in indirect taxes but also received heavy subsidies on inputs.

The low rate of taxation of agriculture was inconsistent with the need for mobilising resources from the agricultural sector. This was especially so in view of the spread of new technology in agriculture resulting in high incomes of surplus farmers, including those with holdings below 8.3 acres. There was a great need for increased investment in agriculture and it was most unlikely that the non-agricultural sector would have been able to generate resources in excess of its own requirements for financing investment in the agricultural sector. Whether land tax or agricultural income tax would have been the appropriate form of direct taxation on agriculture was a moot question. Land revenue was traditionally imposed on plots without reference to the total holdings of a family. For land tax to be graduated it would be necessary for the purposes of tax assessment to pool together all the landholdings of an individual or a family. Moreover, considerations of equity would require account to be taken of differential fertility and differential access to physical infrastructure and modern inputs, including differences in cropping intensity. The collection of reliable information for these purposes would be time-consuming and costly. Therefore, reliance would have had to be placed on rough and ready measures of the differential qualities and productivities of different landholdings. Moreover, revisions in tax rates would have been necessary every few years in response to changes in prices and productivity.

Proposals for the enhancement of the rates of land revenue were made from time to time but did not find much favour with the policy-makers. One such suggestion was to increase the rate of land revenue on holdings above 8.3 acres on a graduated scale starting from Tk.25–45 per acre to Tk.75–135 per acre. This would have imposed a burden of no more than 10% of net income on the biggest farmers. The proposal was rejected by the political leadership on the ostensible grounds that the agriculturalists had suffered heavy losses of income and assets during the war of liberation and that the time was not opportune for imposing land taxes so soon after independence. This did not seem to be valid reasoning; at least a beginning could have been made with low rates, which could have been increased over time. Only in this way could the political leadership have established the proposition that the agricultural sector should not escape the burden of taxation; in fact, the tax burden should have increased with the increase in agricultural income and productivity over time.

In view of the detailed information which was necessary for an equitable and efficient system of land tax, and the possibility that a centrally administered land tax was liable to abuse and corruption on the part of the tax officials, the best way of administering a land tax would have been to delegate it to the local authorities. Whether land was taxed on the basis of land or agricultural income, the local authorities could obtain, with the help of a committee of local representatives, a reasonably accurate estimate of the size of landholdings as well as of yield of land and income from land. The educational cess, the local rate or the development taxes, which were previously assessed in relation to the rates of land revenue, were ideally suited for collection and administration by the local government institutions. If the tax collection machinery itself were at the same time representative institutions, the chances of abuse of power on the part of the tax officials might have been reduced. Only in 1976/77, a beginning was made in increasing taxes in agricultural income; a land tax of Tk.15 per acre was introduced and the agricultural income tax was integrated with the ordinary income tax, even though with a higher exemption limit for agricultural income than for ordinary income.[6]

(ii) *Subsidies.* The mobilisation of resources in the public sector was affected to a significant extent by subsidies provided by the government on outputs and inputs. Under the public distribution system, i.e. the rationing system, foodgrains were sold below cost. The price of rationed food was not only lower than the domestic free market price; it was even lower than the world market price. Over the years, the price of foodgrains distributed under the rationing system was increased in response to the rising price of imported foodgrains but the amount of subsidy continued to be substantial. For example, between 1973/74 and 1974/75 there was a 50% increase in the price of rice and 65% increase in the price of wheat; nevertheless the subsidy amounted to 58% in the case of rice and 38% in the case of wheat in 1974/75. The differential between the price of wheat and rice was reduced over time as consumers in Bangladesh were increasingly getting used to eating wheat. As late as January 1975 the prices of rationed foodgrains compared to the cost of procurement by the government and free market price are shown below.[7]

PRICES OF FOODGRAINS
(*Takas per 2 lb*)

	Rationing Price	*Free Market Price*	*Government Cost*
Rice	1.50	7.75	3.50
Wheat	1.25	1.99	1.99

The magnitude of the subsidy in relation to the budgetary resources of the government was seen from the fact that during 1972/73 and 1973/74 the entire deficit on the revenue (non-development) budget of the government would have been eliminated if the subsidies on foodgrains were abolished. The cost of subsidies were, therefore, met by recourse to deficit financing, i.e. by the creation of bank credit.

The financing of food subsidies involved two additional ramifications with implications for budgetary deficits. One related to the substitution of the domestically-procured foodgrains for the imported foodgrains for distribution under the rationing system. This was the natural consequence of the progress in import substitution in respect of foodgrains. This was indeed the principal objective of food policy. The domestic procurement price, however, if it was to be remunerative to the farmers, could not be out of line with the free market domestic price, which was much higher than the world market price; if the rationed price of foodgrains was to be kept unchanged, unit subsidy had to go up as domestically-procured foodgrains were increasingly distributed under the rationing system. This increased the magnitude of the budget deficit and hence the inflationary pressure. The second aspect of food subsidy related to the amount of counterpart funds which could be generated out of the sale of food aid. Foodgrains imported under food aid were sold at prices below import price; this reduced the amount of counterpart funds generated for financing the public sector development expenditures. The annual loss on this account amounted to about one-fifth of the total annual development expenditures during the years between 1972/73 and 1974/75.

A large part of the subsidised rationed food was distributed to the urban population and government employees, irrespective of the levels of income; moreover, these groups in fact had prior claims on the rationed foodgrains. Whenever there was a shortfall in supply, the requirements of these consumers were regarded as a statutory obligation of the government and were satisfied before the needs of the rural areas or the poorer sections of the semi-urban areas were met. Primarily, rationing in the urban areas with large concentrations of population was justified on the grounds that imperfections in the marketing system would have resulted in very large fluctuations, seasonal as well as annual, in the supply of foodgrains to the urban areas. It could also lead to an unacceptably high differential in prices of foodgrains between the urban and rural areas, especially at times of food shortages, which again could be aggravated by speculative hoarding on the part of both surplus farms and private traders. These were valid reasons for rationing foodgrains in times of overall shortage or of temporary shortages and dislocation of supply. What was the rationale for subsidised food? If the price necessary to stimulate food production was higher than what was within the reach of the poorer sections of the population, then there was a case for supplying subsidised food to the poorer consumers. Ideally, the objective would be to increase the employment and income of the poorer segments of the population so that they were able to purchase food at its supply price. Failing this, the cost of subsidised food to the poorer consumers had to be met by taxes on the consumption of the richer sections of the population. But in Bangladesh neither was the distribution of subsidised food limited to the permanently or temporarily affected segments of the population nor was the subsidy on foodgrains met by taxes on the more affluent segments of the population. The provision of subsidised food to the four large urban centres, irrespective of the economic status of the recipients resulted from the political consideration that a discontented group of the urban middle class and the industrial workers which constituted vocal, volatile and organised pressure groups, would pose a threat to peace and political stability. Similarly, the supply of subsidised food to all categories of government employees all over the country, irrespective of income and wealth, was an attempt to keep this most important instrument for the exercise of state power happy and contented. The rural poor, though in great majority, were unorganised and did not constitute a significant threat to the power structure; therefore, they had to sustain a relative neglect.

If the concern of the rationing system was to provide food at subsidised prices only to the low-paid government workers and industrial workers in both private and public sector industries, then the alternative would have been to abolish subsidies on foodgrains but at the same time to increase the wages or provide income supplements or cost of living allowance to these income groups. Since under this system, the higher income groups would not receive the income supplements or cost of living allowances there would be a saving of costs. Moreover, to the extent that increases in wages or in cost of living allowances would be absorbed by higher prices of goods or services provided by both public and private sector enterprises, there would be no direct burden on the government budget. To some extent, profits of the productive enterprises might absorb the increase in wages. Therefore, while there might be selective increases in prices, the overall inflationary impact of deficit financing, which met the cost of food subsidies, would be moderated. But to the extent that the unorganised rural poor, the landless and deficit farmers had access to subsidised food, its abolition would cause them hardship since they would neither receive income supplements or increase in wages, even though there would be indirect relief from the decline in general inflationary pressure in the economy. In that case, the appropriate policy would have been to provide the poorest sections of the population with subsidised food; the amount of necessary deficit financing would have been less than what was involved in the provision of subsidised food for the entire rationing system. Moreover, food aid should have been exclusively directed towards meeting the needs of the poorest sections of the population.

The cost of subsidies on agricultural inputs constituted a substantial drain on financial resources, particularly subsidies on fertilizers and irrigation water. In 1973/74 fertilizers were sold at less than half the cost of procurement and distribution; as late as 1975/76 the cost of fertilizer subsidies amounted to Tk.800 million and was expected to be nearly as high in 1976/77.[8] The subsidy on irrigation water which took the form of providing water free of charge or at nominal cost was calculated in 1973/74 to range from nearly 50% for shallow tubewells to about 75% for low-lift pumps and as high as 90% for deep tubewells.[9] Thus in the Annual Plan of 1974/75 the annual costs of the irrigation programme for tubewells and low-lift pumps amounted to over Tk.900 million; as against this annual receipts were expected to be about Tk.190 million, resulting in an annual subsidy of about Tk.710 million.[10]

The rate of subsidy on the other agricultural inputs included about 50% for improved seeds and, in the first years after independence, as much as 100% for insecticides and extension service; over the years there was an increase in prices and charges paid for inputs by the farmers, partly in response to rising costs of inputs. The rationale for subsidies on agricultural inputs was well-known; improved seeds, fertilizers, irrigation and pesticides, etc., involved a radical departure from traditional practices. Even though the modern, improved inputs were highly profitable, there was a period of experimentation, education, and training required of the farmers. During this period of "learning by doing" returns from new inputs were lower than their potential yields. Furthermore, there were uncertainties relating to the yield of modern inputs arising from fluctuations in weather, in particular floods and droughts. Not only might the natural calamities wash away the fruits of expenditures on modern inputs; but lack of controlled and regular water supply might further reduce the yields, notably from fertilizer. Moreover, modern inputs were liable to a greater incidence of pests and plant diseases. The loss in crops due to pests and diseases had been variously estimated at between 10% and 25% in Bangladesh. Finally, the administrative failures relating to the procurement and distribution of inputs in the right time, quality and quantity increased the risks of adoption of modern inputs. The presence of these uncertainties considerably enhanced the risks of adopting new methods. Therefore, the prospects of very substantial returns were necessary in order to induce farmers to adopt the new technology. A substantial increase in farmers' income could be obtained not only by the supply of subsidised inputs but also by increasing the price of output. While the price support programmes conferred the benefits of high prices of outputs on all farmers, irrespective of their rate of adoption of new technology, subsidies on inputs concentrated the incentives of high rewards only on the adopters of new techniques.

Bangladesh in fact followed the policy of subsidies on inputs rather than a price support programme. By 1973/74, subsidies had been in existence for more than a decade. During this period there had been a large extension of the use of modern inputs as a result of an increased familiarity with modern inputs, a larger and an assured availability and distribution of inputs and the actual realisation by the farmers of high returns from the use of new inputs. The initial period of "learning by doing" was over. The average use of nutrients per acre of paddy had increased three to four times between 1961/65 and 1971/74 and the area of winter rice crop under irrigation had increased two and a half times. The area under high-yielding rice varieties increased sevenfold between 1969/70 and 1973/74 in the case of Aus; it doubled for Boro, and for Aman the area sown under HYV increased from 0.20% to 9.55% of the total area.[11] Between 1965/66 and 1972/73, imports of pesticides increased from 10,000 tons to 419,000 tons and the supply of the spraying equipment increased correspondingly.[12]

Two propositions in respect of the future expansion of the use of modern inputs came to be widely accepted in Bangladesh. One was that the farmers were now on the whole familiar with the profitability of modern inputs and that they were reaping large profits whenever they could get access to and use modern inputs. The second proposition was that the realisation of the maximum or optimum yields from modern inputs depended critically, amongst other factors, on the institutional facilities, including the quality of the extension services, on adequate and timely availability of inputs, and on the supply of credit.[13]

That gains in output and yield from the use of fertilizer were substantial was evident from the fact that one maund of fertilizer typically yielded an additional output of at least 3 maunds of rice. The value of incremental output of rice resulting from an application of one unit of fertilizer was about four times the cost of fertilizer in 1969–70; as a result of the increasing price of rice combined with continued subsidies on fertilizer, the ratio of incremental output to the cost of fertilizer input increased to six times by 1973/74. In view of a large gap between the cost of and returns from fertilizer, there was considerable scope for raising its price without substantially affecting the profitability of its application.

As in the case of fertilizer so also in the case of irrigation, the gains reaped by the farmers from the use of water in agriculture were very substantial. The returns from investment in water use had been variously estimated between 38% and 85%. As of 1973/74, the increase in the net value of product per acre was eight times the annual rentals paid by the farmers for the use of low-lift pumps; the ratio was 6.6 times in the case of shallow tubewells, 5.1 in the case of deep tubewells and 4.3 in the case of a large-scale irrigation project.[14] Even though such high returns might not always have been realised owing to the organisational and institutional bottlenecks, a three- to fourfold increase in water charges per acre to about Tk.150 would have amounted to no more than 7% of the incremental output or benefit.

By the early 1970s high subsidies on agricultural inputs were unnecessary for the spread of new technology because the farmers were already familiar with and reaped very high returns from inputs, even without subsidies. On the other hand, their continuation resulted in a substantial loss of revenue. There were two more important arguments in addition to loss of revenue against the continuation of subsidies: one was that it encouraged inefficient and uneconomic use of inputs; the other was that it worsened income distribution in the rural areas. That in the presence of high returns and very low cost to the farmers, there was no incentive for the economic use of scarce inputs, was most strikingly demonstrated in the case of pesticides which were supplied free until 1974/75. There was no accountability for the use of pesticides; while there was a shortage of pesticides in some areas there were accumulated surpluses with a deteriorating quality in other areas. The spray operators in order to reduce their work load sometimes gave this equipment along with pesticides to the farmers, especially the influential members of the union council (local government) or to the village leaders, who had little knowledge about their efficient operation and maintenance.[15] The pesticide sprayers had in practice a life of one to two years as compared to their potential life of five to six years. There was also the case of subsidised rice seed which was used for direct consumption rather than as seed because the price of seed was lower than the price of paddy. This was partly facilitated by the inefficiency and delays in the procurement and shipment of imports as well as their transportation and distribution within the country. If seeds were not available in time for sowing, the next best alternative was to use it for consumption, especially when it was cheaper than paddy.[16]

That the benefits of the subsidies were not always received by the intended beneficiaries, especially the small farmers with little or no influence in the distribution and marketing system, was often revealed in various investigations. Because of the scarcity of inputs like fertilizer, prices which the farmers paid were often higher than the officially determined prices.[17] The fertilizer distribution system involved the Bangladesh Agricultural Development Corporation which was a government agency undertaking procurement, transportation and storage up to Thana level (local administrative unit) while the retail distribution was handled by government-appointed dealers or intermediaries as well as by the rural co-operatives. This system resulted in a considerable misallocation both because of inefficiencies and because of the private exploitation of the great gains involved in diverting the sales from the official distribution channel. For example, it was noted in one instance that half of the stipulated allotment never reached a village; the black market prices ranged anywhere from 50% to 100% above the official price.[18] The private dealers naturally took advantage of the scarcity in the market. Also, given the low official price of fertilizers and the overvalued exchange rate and scarcity of essential goods, there was a strong incentive for fertilizer to be smuggled out to India where fertilizer prices were higher.

The benefits of subsidised inputs were not equitably distributed among the farmers of various sizes and income levels. Because of unequal access to the modern inputs owing in part to the organisational and administrative mechanism over which the richer farmers exercised greater influence, and in part to the skewed distribution of credit in favour of the more affluent farmers, the benefits of subsidised inputs were reaped disproportionately by those who from the income distributional point of view needed the subsidy least.

The farmers' co-operatives were considered an important vehicle for the diffusion of modern technology. The modern inputs were channelled through them wherever possible; moreover, the IRDP co-operatives provided training and extension services as well. Among the members of the co-operative societies there was a higher proportion of larger farms than among the non-members. A study in the northern part of Bangladesh revealed that 56% of the members of the co-operatives had farm sizes greater than 2.5 acres whereas 34% of the non-members had farm sizes higher than 2.5 acres. Similarly, 47% of the members and 22% of the non-members had an annual income above Tk.6000. More significantly, the co-operatives were led and dominated by richer farmers; 60% of the members of the Executive Committee of the co-operatives had incomes higher than Tk.6,000.[19]

The supply of credit which was the critical determinant of the extent of use of modern inputs was also greater for the bigger farmers; this was especially true of the supply of institutional credit.

A recent survey in the selected areas of Bangladesh indicated that farmers with a less than average size of holding, which covered about two-thirds of the total holdings, received well under half of the institutional credit, whereas their share of the non-institutional credit was a little over one half. The terms of credit for non-institutional credit were harder than that for the institutional credit.[20] The institutional credit met no more than one-sixth of the credit needs of the farmers and no more than one-fifth of the farms were covered by the supply of institutional credit. The system of granting loans on the security of land placed the tenants and sharecroppers at a big disadvantage in securing loans, except on very high terms. Frequently, there were examples of co-operatives in which the bigger farmers and the members of the managing committee obtained the bulk of the loans.[21] In the case of the Agricultural Bank, about 70% of the loans were provided to farmers with more than 3 acres, and about 30% of the loans went to those with 12.5 acres and more.[22]

That the larger farmers reaped a large share of the benefits of subsidised inputs was also evidenced from the available data on the distribution of the use of inputs by farm size. The per acre use of fertilizer was 20% to 30% higher on larger farms.[23] The low-lift irrigation programme was mainly extended through the organisation of groups of farmers called "pump groups" as well as through the farmers' co-operatives. The bigger farmers dominated the pump groups or co-operatives and had larger access to irrigation water. About 63% of the farmers in these pump groups had holdings of more than 2 acres. Moreover, the managers and the model farmers who were the leaders of the pump groups and co-operatives had larger landholdings than the ordinary farmers. The average area irrigated was also positively correlated with the size of holdings.[24]

In the light of the above circumstances, the arguments for the substantial reduction or elimination of the subsidy on inputs were strong. The modern technology was in any event highly profitable. Frequently subsidies were not received by the intended beneficiaries but were diverted to illegal export or trade; to the extent that the farmers received them, it was more likely than not the rich and surplus farmers. The resources for providing subsidies were derived either from a system of taxation which, on the whole, imposed a heavy burden on the poor or from deficit financing which was highly regressive in its effects. The Five-Year Plan unequivocally recommended the abolition of subsidies over time. Even though in response to the rising cost of procurement of inputs, there was an increase in price, it did not reduce the percentage of subsidy; it merely prevented a rise in the total cost of the subsidy.

However, the policy-makers, while increasingly becoming convinced about the need for reduction of subsidies, were against a substantial jump in prices of agricultural inputs or an abrupt reduction in subsidy. They were persuaded to reduce subsidies primarily because of the paucity of budgetary resources, but they were reluctant converts. This was due to a number of factors. The beneficiaries of the system were the surplus farmers and the intermediaries who in the process of distribution secured a share in the scarcity margins and they constituted an important pressure group, important in both national and local politics. This was most clearly demonstrated in 1973/74 when the Cabinet first decided to reduce the subsidy on fertilizer substantially, but then after the pressure groups in the ruling party had exerted themselves, in large part reversed its decision. Moreover, the policy-makers expressed an apprehension that in spite of the high profits from the use of modern inputs, reductions of subsidy would discourage the rapid expansion of their use. Increasing food production was a high priority policy objective; any deceleration in the spread of modern technology which could be ascribed to the elimination or rapid reduction in subsidy was not acceptable to them. They were averse to taking any risks in this respect. For the proponents of a policy of subsidy reduction, there was no clear way in which to demonstrate that the reduction in subsidies would not delay the introduction or reduce the use of modern inputs in agriculture. One complicating factor in this respect was the fact that there was a shortage of fertilizer, due to both the shortage of foreign exchange for imports and the slow recovery in domestic production; there had been virtually no increase in the consumption of fertilizer between 1972/73 and 1974/75. It was, therefore, decided to allow only a slow rise in the price of fertilizer.

It was also argued that a reduction of subsidies would adversely affect the smaller and medium farms. While admitting that the benefits were reaped more by the large farms, it was pointed out that also the small farmers did obtain a share, even though a relatively small one. To the extent that they would be deprived of the benefit of the subsidy, this would discourage production and also make at least some poor farmers worse off. The government responded to the criticism about the inequity in the distribution system by tightening up controls as well as by introducing various rules and regulations so as to increase the flow of modern inputs to the poor farmers.[25] Moreover, it was argued that the process of elimination of the subsidy should be associated with an increase in the supply of modern inputs. This would keep the prices to the farmers rising less rapidly than the rate of reduction in the subsidy. Similarly, while admitting that a more liberal supply of credit would offset the disadvantages of the reduction in subsidies, they were unwilling to sacrifice whatever benefits were derived from the subsidies, until the credit and distribution system could be considerably improved. In other words, improvements in the associated measures for the spread of modern technology should be undertaken simultaneously with the reduction of subsidies.

There was an increase in the price of agricultural inputs over the years between 1972/73 and 1976/77. Between 1973/74 and 1976/77, as shown below, the price of urea had doubled.[26]

FERTILIZER SUBSIDY
(*Tk. per maund*)

Type of Fertilizer	*1973/74* Cost	*1973/74* Sale Price	*1974/75* Cost	*1974/75* Sale Price	*1975/76* Sale Price	*1976/77* Sale Price
Urea	40.77	30.00	58.00	40.00	50.00	60.00
TSP	56.16	20.00	115.00	30.00	40.00	48.00
MP	42.56	15.00	70.00	20.00	30.00	40.00

However, during 1974/75 for the first time, the pricing system for pesticides was introduced, which nonetheless involved a subsidy of about 50%, compared with cost price of 1973/74. The pricing system, however, did minimise the wastage of pesticides and instituted a mechanism of accountability which was essential for the efficient use of this input. There was a fall in the use of pesticides at the same time that the pricing system was introduced. To what extent it was a reflection of the elimination of wastages of illegal export rather than a reflection of a decline in its effective use was difficult to determine and especially so since, unlike in the case of fertilizer, the institutional arrangements and the training of farmers in the use of pesticides were considerably less developed. In the case of irrigation water, the subsidy was reduced in 1974/75 by doubling the rental charges on low-lift pumps and shallow tubewells and more than trebling the rental of deep tubewells. However, the rate of subsidy continued to be very substantial; the payment per acre on account of rental charges of low-lift irrigation increased by no more than Tk.12 to Tk.20, if the increase in the cost of fuel and oil and additional payment on account of wages of drivers which were formerly supplied free, were ignored. Similarly, the payment per acre for deep tubewell irrigation, on account of rental charges only, increased by no more than Tk.7 to Tk.9. The rate of the subsidy continued to stay higher for deep tubewells than for pumps and shallow tubewells in spite of the large cost and higher capacity of deep tubewells.[27]

(iii) *Mobilisation of Surplus Labour.* The opportunities for the productive utilisation of labour in the rural areas, outside of direct agricultural operations, were many and various, specially in the sphere of capital construction works, i.e. building up of physical infrastructure. Traditionally capital construction projects were called rural public works projects in Bangladesh. They were administered by the local government institutions. The projects were planned and executed by the projects committee of the local government institutions. They did not rely on the contractors; the engineers of the government departments were to help design the projects and provide assistance in supervision. Technical experts from irrigation and road departments were called into service. The rural works programme suffered a serious setback in the years immediately preceding independence. This was partly due to the collapse of the local government institutions, which towards the end of Pakistani rule became an instrument of political patronage and were identified with the Pakistani regime; hence it shared the general discredit to which the Pakistani regime was subjected. The programme was dependent exclusively on food aid and on the generation of counterpart funds. There was no system of direct financing of the rural works programme out of the country's internal budgetary resources. Food aid expanded the supply of wage goods, i.e. food, on which wages of labour employed in the labour intensive public works projects were mainly spent. While in a few projects, financed by the FAO Food for Work programme, wages in kind, in the form of food, were paid, most labourers engaged in the works programme were paid money wages at the going market rates, in the knowledge that demand generated by additional income would be spent overwhelmingly on food.

The problem of utilising surplus rural labour without the payment of the market rate of remuneration engaged much attention in Bangladesh. Given the motivation and opportunity for work, it was considered possible to utilise the idle labour force without payment of market rates of wages, since they were somehow able to obtain "subsistence", partly by intermittent work and partly by dependence on friends, relatives and general charity. What was lacking, therefore, was political leadership, able and willing to motivate and organise the idle labour force, for nation-building activities. However, it was at the same time appreciated that there were several limitations to this method of utilisation of idle labour. First, the projects on which the idle labour was put to work did not always confer direct benefits on the workers themselves. The construction or irrigation ditches or roads or drainage schemes helped those who were landowners and often those amongst landowners who were located near these development projects. The villagers most likely to have free time for local construction activities were precisely those who did not have much of their own land to till—landless labourers and small farmers. To get them to work on projects which improved the land of others, it was only to be expected that they would expect the going rural wage. The private productivity gains from rural projects were, however, seldom large enough for the beneficiaries to pay the wages out of their own pockets. Secondly, friends and relatives or members of the family who shared consumption with the unemployed usually were reluctant to fund or provide subsistence once the idle persons found employment elsewhere. Thirdly, when idle labour was put to work, its consumption went up as well. At low levels of income, there was a positive and strong correlation between work and consumption requirements, especially food consumption. Unless there was a supply of wage goods from external sources, those who were already employed had not only to share consumption with the newly employed but also they had to reduce their own consumption in order to allow an increase in the consumption of those who were newly employed. There were two ways of accomplishing this. The first possibility was to impose on the previously employed increased taxation in order to pay the newly employed on the assumption that taxation would cause the former to reduce the consumption of wage goods. To the extent that these two flows did not match, there would be a need for an external injection of wage goods. Secondly, the newly employed could be paid a lower wage rate, i.e. lower than the average income of those previously employed. It was not always possible to build up such a dual structure of incomes and wages between the old and newly employed labour force.

The Five-Year Plan of Bangladesh considered the possibility of mobilising surplus labour in rural development projects by requiring every able-bodied male adult, irrespective of wealth and status, to contribute a certain amount of physical labour every year. This was equivalent to a "head tax in kind", i.e. in terms of labour input, but provided an option for anyone to commute part of the compulsory labour input, by making equivalent financial contribution, converted at the going wage rate. It was possible to impart progressivity to the "tax in kind" by requiring a minimum amount of man-days of labour per year on the part of every able-bodied individual and by imposing a surcharge in the form of an additional amount of "mandays" payable by those whose income or assets exceeded a minimum level, either in kind in terms of physical labour or commutable in terms of financial payment. The imposition of the "head tax" as well as the utilisation of the man-days contribution as a head tax in the rural development projects was to be the responsibility of the local government institutions.

Assuming that 50% of the adult male labour force would be liable to pay head tax, i.e. about 20 million individuals, and that each individual would contribute at least 8 days of labour a year, 160 million man-days of work would be available. If three days were fixed as compulsory labour requirements for everyone, then 60 million man-days would be contributed in "kind" and partly in financial contribution, depending on the choice exercised by the taxpayers, having more than a minimum level of income or assets. If 90% of the additional man-days are paid in financial terms, this would amount to about Tk.600 million at the going wage rate of Tk.6 per day. This was a substantial amount of financial resource, to be used in employing the landless labourers or farmers owning very little land and having a considerable amount of idle labour. This method of mobilising labour was tantamount to taxing the more affluent and relatively "less unemployed" labour force to finance the employment of landless or employed labour force.

There was one group in society, however, which could without any extra remuneration play a dominant role, with their high spirits and enthusiasm, in organising the idle unskilled labour force. They were the students who could provide the managerial and skilled labour input and could work as extension agents, especially in selecting and supervising projects as well as making a contribution to manual labour. They could also help in determining the minimum level of income or assets above which "surcharge", i.e. additional tax was to be imposed, and in identifying those who were above the minimum level through open discussion and exchange of information regarding income and assets (an important requirement in the absence of records of income or assets in the villages) among the rural population; they could also assist in imposing and collecting the appropriate amount of surcharge. Provided their vacations were rearranged to coincide with the timing of the planning and the implementation of projects they could provide a continuous input of 2–3 months of labour.

The system of labour utilisation described above required an efficient local government under a forward-working political leadership so that privileged groups in rural society with high income and wealth did not subvert the system. These groups would be opposed to the compulsory input of physical labour as well as the surcharge of additional tax, of which they had to bear the full burden. Such a mobilisation of the rural labour force would require not only an efficient self-governing institution at the "grassroots" level; it would also require a more egalitarian distribution of income and assets, especially land.

While the ways of mobilising surplus labour in rural development projects were discussed in the Five-Year Plan, they were not taken up as a policy measure at an operational level by the government. In early 1972 there was some thought given to the idea of using the freedom fighters (Mukhti-Bahini) who had a great deal of zeal and enthusiasm for public service and who could be trained and mobilised for organising the rural masses for productive work. This was also the period when a certain amount of thought was given to the question of redistribution of assets, especially land ownership. However, none of these possibilities were exploited or were seriously explored by the political leadership. Moreover, the development of the self-governing rural institutions was delayed inordinately. As has already been pointed out earlier, any successful attempt at the mobilisation of surplus labour would also require or would be greatly facilitated by active participation by the labourers themselves in the selection and administration of development projects.

It was obvious that for such a policy to succeed one needed a determined and mobilised political party with cadres throughout the countryside. It was unlikely, however, that the ruling party in Bangladesh could integrate the Mukhti-Bahini as a part of its cadres without a vigorous and determined push from the top. There were differences in ideology and outlook between groups of freedom-fighters as well as between the freedom-fighters, on the one hand, and the ordinary members or the rank and file of the ruling party, on the other.

In the early years after independence, in the context of a widespread illegal possession of arms, amongst the rural population, centralisation rather than decentralisation was the corner stone of the government's policy with a view to controlling the law and order situation and with a view to bringing illegally held arms and weapons under the government's control. Decentralisation of administrative power ran the risk of chaos and confusion at the level of the local government, if arms could not be recovered from the private hands. Gradually, however, there was an improvement in law and order as well as a substantial recovery of arms. But there developed a resistance to the devolution of administrative and financial powers to the local governments on the part of both the politicians and administrators at the centre. The latter disliked the loss of power and the inability to distribute patronage. The former dreaded the possibility of opposition political parties or opponents in the same political party gaining control over the local self-government institutions.

The fact that this method of resource mobilisation was not seriously considered for implementation in the past did not imply that this was an unimportant or a dead issue and that this would not require renewed attention or acquire importance in the future. Now and for many years in the future, manpower would remain the greatest or most important resource of Bangladesh and, therefore, the question of how to mobilise for productive work in the most labour-intensive way and with a minimum increase in the supply of wage goods would continue to plague the policy-makers in Bangladesh. With an increasing population, the man–land ratio is going to worsen in Bangladesh, in future years; open and disguised unemployment and an increasing number of landless labourers would dominate the economic landscape of Bangladesh. Increasingly, it would be difficult to provide employment for a growing labour force within the market economy system under the conventional system of wage employment. A continued infusion of external resources from the rich countries to provide an elastic supply of wage goods, from abroad should be considered an unrealistic possibility. Similarly, a spectacular breakthrough in stemming the growth of population was ruled out in the foreseeable future. Even if the output of food, i.e. principal wage goods, was to be expanded considerably, how to mobilise it from those who produced it for the benefit of those who did not, without serious discentive for increased productivity under a system of market economy and under a system of private, individual ownership of land and assets was going to pose a dilemma and to constitute an important policy issue for Bangladesh. Therefore, it might be necessary to think of work-sharing arrangements for the rural population.[28]

That domestic resource mobilisation in Bangladesh in view of her general level of poverty and her low capital income would remain a very difficult uphill task was and will be readily foreseen. However, there were and will be pockets of savings or surplus in a general environment of poverty and deprivation, which could and should be mobilised for investment in the public and private sectors. The possible sources of capital accumulation, emphasised in the preceding pages, were mainly profits in the trading and industrial sectors, both public and private, and high returns or incomes accruing to the surplus farms from the spread of new technology. It should also be noted that the government's policy to impose sacrifices on current consumption through taxation and related measures should have been not only too easy to implement but also was likely to achieve credibility and political acceptability if the burdens of sacrifices in current consumptions were equitably shared between the rich and the poor, and if ostentatious consumption and levels of living on the part of the rich were restrained. Moreover, to the extent that the generation of savings was linked with the opportunities of profitable investments on the part of the savers, it was necessary to stimulate or expand opportunities for private investment. This was especially true if financial mechanisms, i.e. financial assets and institutions, were inadequate to mobilise private savings to the desired extent. To mobilise small savings, specially in the rural sector and the informal urban sector, it would be essential to open up opportunities of investment for the small investor in such areas as marketing, distribution, transport, small-scale and household industry, fishing, agricultural implements and irrigation equipment, etc. It was necessary in the prevailing circumstances of Bangladesh to seek a delineation of investment opportunities between the private and public sectors which was consistent with the proposed contribution of private and public savings to capital accumulation.

If private voluntary savings which could be mobilised in the form of financial assets to finance public investment fell short of the requirements, it was necessary to seek alternative ways of mobilising resources for the public sector through taxation; for example, a disproportionate dependence of public investment on the mobilisation of voluntary private savings through the financial mechanism might frustrate the achievement of both the investment and saving targets. The requirements of investment in the agricultural sector were vast in view of the need to employ a large and increasing labour force. It was unlikely that in the foreseeable future the non-agricultural sector would be able to transfer resources for agricultural investments; in fact, as income and output growth in the agricultural sector gathered momentum, it would be the latter which would be required to finance investment in non-agricultural activities predominantly within the rural economy rather than in the urban sector. Substantial resources were to be mobilised from the agricultural economy for agricultural and rural investment. Local resources were to be found for investment in local educational and health facilities, in local roads, in marketing and distribution channels, in the provision of irrigation, drainage and water supply, etc. It was not only that it would be necessary to organise local institutions to plan and implement as well as to find local resources for a wide variety of rural projects; but also it might be necessary to make investment by the central or national government, including that of foreign economic assistance, in rural development projects conditional upon the contribution by the local community towards their financing either in cash or in the form of supply of land, labour and construction materials, etc. This form of linking the financial contribution by the central government with investment resources to be mobilised by the local government institutions especially for projects which meet the felt needs of the local community was not new and could be effective in mobilising local resources for investment.

Notes

1. J. Faaland and J. R. Parkinson, *Bangladesh: The Test Case for Development,* C. Hurst and Co., London, 1976, pp. 79–80. Net inflow of resources to Bangladesh is estimated to have exceeded the total invested during the early years of Bangladesh.
2. The Planning Commission, *Five-Year Plan,* pp. 49–50.
3. J. Faaland and J. R. Parkinson, op. cit., pp. 50–2, 82–3.
4. *Annual Budget,* 1974/75; *Budget Speech,* 24 June 1974.
5. Planning Commission, unpublished memo on a Proposal for Increase in Land Revenue, 1973/74, June 1973.
6. Ministry of Finance, *Annual Budget Speech,* June 1976.
7. Government of Bangladesh, *Ministry of Food and Civil Supplies.*
8. Ministry of Finance, *Budget Speech,* 1976/77, p. 26.
9. Planning Commission, unpublished memorandum.
10. ibid., unpublished memorandum; Planning Commission, *Economic Review,* 1974/75, pp. 45–6.
11. R. Ahmed, *Foodgrain Production in Bangladesh: An analysis of Growth. Its Sources and Related Policies,* International Food Policy Research Institute, May 1976, appendices, pp. 331–7, 339.
12. M. Rahman, "Economic Incentives and Rural Income Distribution: Some Observations in the Light of Bangladesh Experience", unpublished paper, *Bangladesh Planning Commission,* 1973.
13. For example, an analysis of the inter-district variation in the use of fertilizer revealed that topography, farm size, transport, communications, storage facilities, industrial production, trade and commerce, etc., all played a part. The most important determining factors were identified as (*a*) relative flood hazards, (*b*) transport and communications, (*c*) proportion of owner-operated farms, (*d*) extent of on-farm activities, (*e*) extent of irrigated area and (*f*) credit availability and quantity of investible surplus on the farm.
14. *Five Year Plan,* p. 150.
15. M. Rahman, op. cit. Private and powerful business groups exerted indirect influence through attractive terms of sale to inflate the requirements of imports which were not actually used but illegally re-exported or wasted.
16. ibid.
17. The farm level fertilizer prices were often 44% higher than government fixed prices. R. Ahmed, op. cit., p. 16.
18. M. Rahman, op. cit.
19. The above must be appraised in the context of distribution of landholdings in Bangladesh where about 60% of the farmers had holdings of 2.5 acres or less. I. Ahmed, "The Green Revolution in Bangladesh: Adaptation, Diffusion and Distributional Questions", *Seminar on Socio-Economic Implications of Introducing HYVS in Bangladesh,* Bangladesh Academy of Rural Development, Comilla, Bangladesh, p. 25.
20. ibid., Tables A–1 and A–2.
21. In Comilla 24% of the members received 63% of the loans and 69% of the overdue loans were owed by the members of the managing committee. M. Rahman, op. cit., p. 24.

22. IBRD, *Bangladesh Agricultural Credit Review*, June 1974, Appx. II.

23. M. Muqtada, "The Seed Fertilizer Technology and Surplus Labour in Bangladesh Agriculture", *The Bangladesh Development Studies*, October 1975.

24. M. Alam, "The Beneficiaries of HYV under Thana Irrigation Programme in Bangladesh", *Internation Seminar*, 1975, Bangladesh Academy of Rural Development, Comilla, Bangladesh, p. 17.

25. For example, during 1974/75 a new procedure was introduced for the distribution of fertilizer under which the ministry of agriculture was to make allocation to individual districts; allocation to different thanas within each district was made by district committees consisting of officials and non-officials. Allocation to different unions in each thana was made by the thana committee and allocation to private dealers at the level of unions was made by the union committee headed by the Chairman of the Union Council, i.e. the local government unit at the union level. The committee also decided on the list of farmers who were to be provided with fertilizer by the dealers. *Economic Review*, 1974/75, op. cit., p. 35.

26. Planning Commission, *Economic Review*, 1974/75, June 1975, pp. 52–4. Also *Budget Speech*, 1976/77.

27. Since subsidies were fixed not in terms of a certain percentage of the cost price of inputs but in terms of an absolute differential between cost and sales price of inputs, the impact of a rise in the sales price on a percentage rate of subsidy in a general inflationary situation, depended upon the increase in the sales price in relation to the increase in cost.

28. J. Faaland and J. R. Parkinson, op. cit., p. 147–9.

7. Industrial Policies: Public versus Private Sector

(a) Introduction

The evolution of investment policy with respect to the relative roles of public and private sectors in the field of industry illustrates an interesting interplay of economic, political and ideological factors. The industrial sector became not only the test case for the fulfilment of important electoral pledges but specifically was seen by many as a first major step towards fulfilling the commitment to a socialist transformation of the economy. It revealed both the possibilities and the limitations in the way of extension of public ownership; it became a symbol for the struggle between the contending socio-political forces far out of proportion to the importance of the industrial sector in the economy. The pressures and counter-pressures generated by the extension of public ownership in this sector brought out in the open all the problems and contradictions, discussed in the earlier chapters, in the socio-political milieu of Bangladesh. It was also in this sector that the issues of incomes policy as well as of concentration of economic power were debated at length. These issues concerned not only the structure of incomes and incentives for the managers of the nationalised industries but also the level and structure of individual wages, the role and functions of trade unions, including the question of their participation in the management and profits of public enterprises. And it was in this sector that the role of foreign capital and technology transfer became an important issue. This chapter discusses each of these issues in turn.

(b) Nationalisation Measures

In March 1972 the government took over public ownership all industrial units abandoned by the Pakistanis with assets above Tk.1.5 millions; furthermore, while promising to pay compensation, it nationalised all units owned also by Bangladesh citizens in three large-scale industries, i.e. cotton textiles (excepting handloom and specialised textiles), jute manufacturing and sugar manufacturing. Before independence, cotton textile, jute and sugar mills were partly owned by Bangladesh citizens, either singly or jointly with the public sector, and partly by Pakistanis; the proportion of total assets owned by Bangladesh nationals in the jute and cotton textile industries was 34% and 53% respectively. In other industries Bangladesh private enterprise—which was anyway insignificant compared with that of Pakistanis—was unaffected. Before independence, total industrial assets under Pakistani ownership constituted 43% of all industrial assets under private ownership; of this about 83% was put under public ownership. As a result of the nationalisation, the share of the public ownership in total industrial assets went up from 36% before independence to 86%.[1]

It is a moot question whether Bangladesh so early after independence would have embarked upon such a significant nationalisation policy if she had not automatically gained public ownership of a large part of the industrial sector by simply taking over the units left behind by Pakistanis; in any case it acted as a catalytic element. Thus, in the government's major statement on industrial investment policy in March 1972 the Preamble makes special reference to this fact:

> As an immediate and direct result of the liberation war, the predatory Pakistani industrialists had "abandoned" their industries in Bangladesh but only after taking with them everything they could, including, in some cases, the provident fund money of the employees. They left behind only such plant and machinery which they could not remove and also left behind huge liabilities. In this situation, the government was faced with the immediate task of restoring the existing industries. This meant the assumption by the Government of the enormous responsibility of replacing the entrepreneurs and managers for efficient and effective management in the public sector.[2]

Could the government have pursued the alternative policy of selling out the abandoned industrial units to prospective Bangladesh entrepreneurs? This raised two sets of questions: Was such a policy desirable in the socio-political setting of post-independence Bangladesh? Was this feasible? As has been explained earlier, both within and outside the ruling party, the radical aspirations were built up during the last days of the Pakistani regime and during the war of liberation. A political party which wanted to retain its mass appeal and support amongst the students and workers could not easily disregard the prevailing political mood.

There were only a few industrial entrepreneurs in Bangladesh, mainly concentrated in the jute and cotton textile industries, with experience of managing and financing large industrial entreprises.[3] The large "abandoned" industrial units could only have been transferred to them, if they were to be handed over at all to private enterprises. Not only this, but these individuals would have needed large-scale assistance from the state in the form of loans and credits to be able to purchase such industrial units. Moreover, the abandoned industrial units had suffered from severe depletion of assets, working capital and inventories in the aftermath of war, not to speak of financial liabilities left behind by the Pakistanis. The private investors would only have been persuaded to purchase them if the government was ready to compensate them for past losses and to contribute to their rehabilitation. Given the need for massive help from and intervention by the government to put them in working condition, the merit of handing them to private investors was in serious doubt. It would have resulted in the government actively building up a regime with a few entrepreneurs having ownership and control of the industrial sector. These were precisely the kinds of policies which had been followed in Pakistan. The dominance of a few industrial families in Pakistan and their influence on the economy and politics of Pakistan was already a matter of serious criticism and concern in pre-independence days, particularly in Bangladesh.[4] There was a general scepticism in the country, including within the ruling party, against the repetition of such policies.

That private entrepreneurs would in any case have lacked sufficient interest in taking over the large industrial units seemed subsequently to be confirmed by the experience with the policy of disinvestment of the smaller abandoned industrial units, i.e. units with less than Tk.1.5 million in assets. By the middle of 1974 out of three hundred such small industrial units, a many as two hundred and thirty-two units were yet to be assessed, in terms of their value of assets, prior to transfer to private hands. The progress was slow in spite of the fact that the intending purchasers could pay in instalments, and in some cases, by bank borrowing.[5] In many of the abandoned enterprises, militant workers took over control and management; in a few cases, they firmly resisted transfer to private investors. Such action on the part of workers partly reflected radical expectations fostered by promises of the political parties in the past; partly it was an act of indiscipline aggravated by a spirit of revenge against the former Pakistani owners or managers and their collaborators—forgetful of the fact that they were no longer relevant in independent Bangladesh.

It is interesting to recount the reasons which were advanced for the slow progress in the transfer of the small, abandoned industrial units to the private sector in Bangladesh.[6] Many of these reasons would have applied with added force to the case of the larger units. There was a general sense of insecurity among the owners caused by widespread private possession of arms, procured during the war of liberation; it dampened the investment climate and created uncertainty in the minds of potential investors about the security of private capital or the certainty of returns from investment.

Frequently people with surplus funds, legally or illegally earned, did not have entrepreneurial ability or interests; they were interested in quick turnover and prospects of high profits and, therefore, mainly in trading and speculation ventures. Again, since many units were saddled with machinery which had not been properly maintained or repaired, the potential investors were apprehensive about the quality of equipment and machinery. Moreover, in some cases small units were previously operated by the Pakistani self-employed skilled workers; the potential investors in Bangladesh had neither the skill nor willingness to assume the role of self-employed workers.

The extension of public ownership to the abandoned industrial units provided an occasion for a serious consideration of the electoral pledges given by the ruling party that heavy or basic industries including jute, cotton textiles and sugar industries would be nationalised. Heavy industries such as iron and steel, fertilizer, heavy chemicals, pharmaceuticals, newsprint, etc., were already placed under public ownership before the emergence of Bangladesh. In the context of pre-independence Bangladesh, the jute, textile and sugar industries were also considered to be the commanding heights of the industrial economy and thus were eligible for public ownership. In the early days of united Pakistan they had special relevance since it was through their ownership and control that the Pakistani industrialists obtained a hold over the economy of Bangladesh. In a way, the nationalisation of these industries was considered in those days an important means of liberating the vital sectors of the Bangladesh economy from Pakistani economic domination. The jute manufacturing industry was the largest industry in terms of employment of labour and capital as well as contribution to foreign exchange earnings. It was closely related to the production and trade in raw jute. The electoral pledge, included the nationalisation of both internal and external trade in raw jute which was expected to ensure high and stable prices to the growers of raw jute. The electoral pledges before liberation were carried over into the post-liberation period. In the objective circumstances of Bangladesh obtaining in the early days of 1972, the opposition to the policy of nationalisation was weak; the private vested interests and the nascent industrialists were small, ill organised and mostly on the defensive, given their passive role in the liberation struggle. Moreover, the nationalisation of a few Bengali-owned enterprises was not considered a heavy burden on a greatly extended public sector caused by the phenomenon of abandoned properties.

(c) Limits to Private Investment

The nationalisation measures outlined above were accompanied by imposition of limits or ceilings on private investment in July 1972. New private investment was to be allowed in units with assets of no more than Tk.2.5 million, which could grow up to Tk.3.5 million through reinvestment of profits. The public sector was however not debarred from investing in enterprises with assets below Tk.2.5 million. The Bangladesh Small Industries Corporation which was entrusted with the responsibility of promoting small-scale industries with financial and technical assistance, could pioneer small-scale enterprises, if private domestic enterprise was not forthcoming, through direct ownership either alone or jointly with worker's co-operatives or foreign investors. Moreover, the government reserved the right to nationalise any private enterprise which—due to neglect or mismanagement—was consistently making losses or under-utilising its capacity. Subject only to these provisions, there was a moratorium for the next ten years on further nationalisation of private enterprises, either existing or to be established within the Five-Year Plan period.

The basic rationale behind the ceiling was that the private industrialists were not to be allowed to grow into big capitalists through the expansion of the existing enterprises either by reinvestment of profits or by means of external financing. The financial limit of Tk.2.5 million was rather arbitrary, it had some relationship to the criteria by which small-scale industries were defined for purposes of labour legislation and other government promotional or regulatory measures relating to the industrial sector.[7] A further consideration in not setting the ceiling at a lower level was the constraint on the managerial and organisational ability in the public sector. The public sector, having greatly expanded within a short space of time, would have found it difficult to manage a very large number of small enterprises.

The expansion of the assets of private industrial units from Tk.2.5 million to Tk.3.5 million and no more, was considered adequate for meeting their growth objectives. It was presumed that such an expansion, financed mainly by the reinvestment of profits, would take place mainly, if not entirely, in response to technological change and in order to realise economies of scale. Putting the investment ceiling in terms of assets rather than employment was also expected to stimulate labour-intensive techniques of production. While there was a limit on the size of an individual enterprise, there was no limit on the number of units a particular individual or groups of individuals could own. A significant expansion of private industrial fortunes, through the ownership of multiple enterprises, was, however, considered unlikely without substantial financial assistance from the state. In the first place, the complexity and administrative burden of managing a large number of small enterprises, compared to the task of managing the same amount of capital and labour in a single enterprise, was a deterrent. Moreover, since expansion would depend on financial assistance from the state, the latter could regulate the rate of expansion by its credit policy.

While demarcating the respective roles of public and private investment in 1972, the government recognised that there were alternative ways of achieving such a demarcation. It was possible, for example, to reserve a few selected sectors for public investment, especially those requiring large capital, sophisticated technology and managerial expertise, and the rest could be kept open for private investment. Another way of curbing the growth of large private fortunes would have been to require that the expansion of private units beyond a certain limit would be conditional upon the government acquiring a partnership, with or without the majority control by the government. Consequently, a small private enterprise, as it expanded into a large size could only do so by converting itself into a joint enterprise with the state.

These alternatives were rejected by the government, even though there were divergences of opinion in the Cabinet on this issue, which partly reflected the differences of opinion about the future economic system and the role of private enterprise in such a system. Public ownership of major industries was consistent with an economic system in which private enterprise was to play a predominant role; public enterprise could alternatively play a transitional role in the progress of an economy towards a full-fledged private enterprise system. This was the role assigned to it in pre-independence days; it was to fill the void whenever private enterprise was not forthcoming for lack of capital, knowledge, managerial capacity or technology. However, a limited number of public enterprises might continue to prevail in the long run in a private enterprise economy, wherever private costs and benefits diverge widely from each other either because of a long gestation lag, need for a long-term planning horizon, natural monopolies, etc. Viewed in this light, the role of public enterprise in the past would merely be to supplement private enterprise; in fact, as and when private enterprise was to be forthcoming, the public sector enterprises could be handed over to private investors.

In 1972 there was a strong section of opinion in government which visualised the relative roles of public and private enterprise very differently. They conceived of private enterprise as a transitional phenomenon, in the progress towards the socialist transformation of the economy. In this view, to prevent the growth of large-scale industrial bourgeoisie was an essential component of the policy of pre-empting the growth of anti-socialist forces in the future. A few, if not all within the policy-making circles, felt that any opportunity for private and public enterprises to co-exist in the field of large-scale industry, either independently or as joint enterprises, would keep the door open for the pressure of anti-socialist forces, i.e. pressure by private industrialists themselves for the expansion of the scope for private investment. Moreover, private interest groups in the field of industry could in turn become a lobbying force for the expanded role of private enterprise in the other sectors of the economy. Within this vision of the future economic system, the ceiling on the size of private investment was set so as to allow it only limited opportunities in the sector of small-scale and cottage industries.

(d) Private Foreign Investment

The attitude towards domestic private enterprise, as enunciated in the industrial policy statement in 1972, determined in turn the policy towards foreign direct investment. Foreign direct investment was allowed only in collaboration with the public sector corporations and only with a minority equity participation of 49% or below. However, no specific field of activity was excluded from participation by foreign investors; public enterprise could invite foreign collaboration in any field. All acts of foreign collaboration, however, were to be approved by the government. This was basically not different from the system of licensing of domestic private industrial investment. The public corporations were free to enter into management contracts with the foreign enterprises in order to take advantage of their superior managerial, marketing and technical skills. Full freedom was guaranteed for the transfer of annual profits (after the payment of taxes) and for the repatriation of capital spread over a number of years. A minimum dividend of 15%, subject to the availability of profits, was also guaranteed.

The exclusion of foreign private enterprise from partnership with domestic private interests was designed to forestall the possibility of a powerful pressure group being created for the expansion of the private sector. In particular, this might have created pressure for the upward revision of the ceiling on private investment, since small-scale industries were unlikely, in view of their requirements of capital and technology, to attract private foreign investment. It was also believed that there was a relative identity of attitude and ideology, as demonstrated elsewhere in the developing world, between private multinational corporations and the capital-rich countries and institutions. Since Bangladesh was so heavily dependent on liberal economic assistance from rich capitalist countries, pressure for a change in domestic policy could gain added strength. Restricting the association of foreign private investors with the public corporations was expected to contain the influence and reduce the pressure, which might be potentially exercised by foreign interests on the domestic economy and policy, while at the same time allowing Bangladesh to secure access to technology and capital. A large public corporation would also have bargaining power in its negotiation with the foreign investors, which would far exceed that of small, private domestic enterprises competing for or seeking foreign collaboration.

The regulations governing remittance of profits and repatriation of capital were more liberal than in many other developing countries. Thus the provision for a minimum dividend of 15%, subject to the availability of profits, was intended to assure the foreign partners that the majority control by a public corporation could not be used to reduce the distributed profits by diverting them to reserve funds or excessive depreciation allowances or reinvestment and expansion. It was expected that this provision would provide an incentive towards greater productivity and higher profits. There was, however, a risk that since the payment of dividends at 15% on investment was a first claim on any surplus, there might be inadequate provision for reserves or for expansion.[8] Similarly, the provision for management contracts was intended to allow foreign partners a freedom from day-to-day interference in management matters, especially since in public enterprises the possibility of political pressure on the locally-controlled board was not unlikely. Again, guarantees were provided that no nationalisation would take place within a period of ten years, and that fair and equitable compensation would be provided in case of nationalisation. These generally liberal conditions for foreign collaboration were intended to offset whatever disincentives were involved in insisting on the public sector as the only partner with foreign investor.

Subsequent discussions with potential foreign investors revealed that they were less worried about compulsory participation with the public corporations than about the requirement of their majority participation. Exclusion of private domestic investors from joint participation was seen by them, more than anything else, as a reflection of an unfriendly or unfavourable attitude of the Bangladesh government towards private enterprise in general and, therefore, by implication and at least potentially, towards foreign private investment. They therefore pressed for a change in the provision for majority participation by the public corporations. They further argued that in view of its majority ownership, a board of directors would be dominated by a public sector corporation. The management contract, even if it delegated in theory full managerial powers to the foreign partners, would not be able to effectively counterbalance the overriding authority and decision-making power of the board and hence would not be able to exclude political interference entirely.

The domestic private investors, on their side, felt that the willingness of the public sector to accept foreign, but not domestic, private investors as partners, was an inconsistent policy and was an act of discrimination against them. The government's acceptance of its partnership with foreign private investors and not with domestic investors, was justified on the grounds that domestic private investors, unlike the foreign investors, brought no advantages in terms of access to skill, management and technology.

(e) Management of Nationalised Industries

The appropriate institutional framework of and the policy guidelines for the operation of the nationalised industries was a matter of continuing debate and discussion in Bangladesh. The basic question underlying the debate related to the optimum degree of autonomy to be enjoyed by the nationalised industries in their investment and production decisions. What should be the degree of freedom the nationalised enterprises might enjoy, for example, in the hiring and firing of the workers and managers, in selling their products and purchasing their inputs, or in the choice of a product, its price, its techniques or scale of production? At the organisational and institutional levels, there were several issues. For example, should the individual enterprises be grouped together in large conglomerations to be called sector corporations (sector referred to a group of enterprises producing similar or closely related products)? If so, what would be the functions and responsibilities of the sector corporations *vis-à-vis* the Ministry of Industry? What degree of autonomy would they enjoy? What would be the functions and responsibilities of the sector corporations in relation to the individual enterprises?

The nationalised industrial enterprises were initially grouped in ten sector corporations, such as chemicals and fertilizers and iron and steel, newsprint and paper, food and other products, etc. There was some debate as to the optimum number of such corporations. By 1976 the number of sector corporations were reduced by about half. The number of sector corporations was not unrelated to the range of functions and responsibilities with which they were entrusted. In the initial years, there was a bias in favour of concentration of power away from the individual enterprises towards the corporations. It was felt that there was a need for central direction in establishing management structures which were disrupted by the war as well as in setting up an appropriate system of financial accounts. The corporations were also to help individual enterprises in procuring supplies, at home and abroad, and to arrange or provide guarantees for loans and credits from the banking system. There were also economies of scale to be derived by concentrating some of these operations in the hands of corporations.

The most important aspect of the management of the nationalised enterprises was the relationship between the Ministry of Industry and the sector corporations. The chairmen of the sector corporations were appointed by the government. It was expected that the chairmen in turn would have the freedom to choose the rest of the officers of the corporations on terms which would be negotiated in each case, or would be determined in light of the market supply and demand for particular skills needed by a corporation. However, such a delegation of authority was contested by the Ministry of Industry; it wanted the appointments in the corporations to be made not only with the approval of the ministry but also on the same terms and conditions as those of the employees in the ministry or government. At one stage the government went so far as to suggest that the Public Services Commissions which selected the civil servants in the various ministries should also be entrusted with the function of selecting managers, professional and technical personnel in each industry. This suggestion was soon abandoned as being overly restrictive and time consuming; the Ministry of Industry, however, continued to restrict, in a few cases by direct intervention, the freedom of choice of the corporations with respect to the hiring and firing of personnel. In fact, the employees of the former Bangladesh Industrial Development Corporation, which was the only public sector conglomerate in pre-independence days, were arbitrarily allocated to the different corporations by the government without the latter being able to exercise any choice as to their suitability. This was considered by the corporations an initial constraint, to start with, on their autonomy with respect to personnel and employment policy. Moreover, with respect to purchasing and marketing operations, they were obligated, like other government departments, to make all purchases at home or abroad on the basis of open "tender" or "bids" invited from the prospective suppliers. For all purchases beyond Tk.10 million they had to secure the approval of not only the Ministry of Industry but also of a Cabinet sub-committee—especially appointed to scrutinise all purchases by the government agencies and public enterprises. These regulations effectively restricted their opportunities for making advantageous purchases at short notice by negotiations with suppliers at home or abroad. These opportunities were particularly important in years between 1972 and 1975, when world-wide inflation, fluctuating exchange rates, speculation in world commodity markets, created considerable uncertainty with respect to both availability and prices of critical imports. Delays in decision-making and procurement operations meant not only a difference in prices to the extent of 100–200% but also an absolute shortage of inputs or an indefinite interruption of production.

Furthermore, the Ministry of Industry stipulated that the sector corporations should not directly deal with the various administrative ministries in such matters as the allocation of import licences or foreign exchange or of annual budgetary allocations from development budgets. On behalf of and in response to requests from the sector corporations the Ministry of Industry was to deal with the relevant ministries with respect to each of these matters. That the ministry chose to intervene or act as an intermediary in such numerous day-to-day issues for which it had neither the competence nor the manpower and time resulted in bottlenecks and delays.

The intervention by the Ministry of Industry or the Cabinet in the detailed operations of the enterprises was justified on several grounds, especially in the initial years. During the interval between December 1971, when the Pakistanis left, and the middle of 1972, when the sector corporations were established, nationalised enterprises were managed by administrators who were often appointed on political considerations. Resultant mismanagement or dislocations led to the need to remove many of these administrators, initially appointed in 1972, and to replace them by more professionally-oriented managers and experienced traders. In this important but politically difficult task, the corporations found the intervention or intercession by the political authorities to be necessary and helpful, although in a few cases it worked in the opposite direction by introducing political considerations in new appointments.

There was the need to deal with competing labour unions, linked up with conflicting political parties, as well as the need to contain the demands for increased wages and improved conditions of work. In this respect there were at least two issues which had political ramifications. One was the problem of the excess labour force, which grew out of the appointment of additional labour during the war of liberation to replace those who either fled during the war to their village homes or joined the guerrilla forces. The old workers returned after the war and the new workers could not always be removed. Also, there was the problem of choosing between temporary, casual workers, on the one hand, and long-term regular workers, on the other. The distribution of the excess labour force between industries and regions was not evenly balanced so that a reallocation of workers between them could have provided some relief. All these interrelated issues required delicate negotiations with labour unions and intervention by the political authorities. Second, there was the problem of payment of arrears or past wages and salaries for the period of the war of liberation for those workers who had abstained from work either for fear of the occupation army or because of their engagement in the war. In many cases the labourers forced the management to pay arrear wages and salaries, not only by going on strike but also under the threat of physical violence. Temporary suspension or withdrawal of authority from the corporation in matters of personnel or employment policy or in matters of negotiations of wage claims, and their concentration in the hands of the government could spare the management of enterprises the threat of physical violence and frequent strikes.

There was an additional problem relating to the financial liabilities of the abandoned enterprises, which also required government intervention. These liabilities were undertaken by the Pakistani owners during the last days of the liberation war, from the banks many of which were owned by them in a system of vertical integration; most of these liabilities were not backed by corresponding assets. This was a problem which the individual corporations could not resolve with their bankers on the basis of the client–banker relationship. There was the need for a general governmental policy regarding these liabilities, i.e. whether to write them off or to convert them to long-term liabilities and if so, on what terms and conditions.

Apart from the need for government intervention to solve specific problems which were peculiar to the post-independence circumstances, there was also the uncertainty in the minds of political authorities about the consequences of autonomy on the part of corporations. Basically, it was the apprehension that there would be corruption and misuse of power which needed to be checked by a close watch and control on their operations. For example, the chairmen of corporations could overspend on overhead expenditures or pay extravagant salaries or wages to themselves and their workers. That close day-to-day control by the government could be counterproductive by sapping initiative, enterprise and sense of responsibility of corporations was not readily recognised; it also made it difficult to enforce accountability or fix responsibility. What was lacking was a set of "rules of business", which would have delineated the powers and responsibilities of both the ministry or the government and the corporations; the rules of business were not finalised until late 1975. This was partly because of the conflict between the ministry and the bureaucracy, on the one hand, and the corporations, on the other; while the former sought greater direction and control, the latter pressed for autonomy as well as higher status and authority. The devolution of the financial and executive autonomy to the corporations was complicated by the fact that the chairmen of corporations who were mostly professional managers, claimed status and authority similar to the highest cadre of the permanent civil servants and, furthermore, wanted to operate independently of them and deal directly with the ministers for the purposes of receiving instructions or rendering accounts of their work. But permanent bureaucracy was not quite ready to share with the professional managers the status and authority which in the past they had traditionally monopolised.

The Five-Year Plan advocated a decentralised administrative structure for the nationalised industries; it conceived of the ministry of industry as performing the task of co-ordination and evaluation. It was to set the goals and criteria of performance for the nationalised enterprises, to gather detailed information about their operations, both current and investment activities, and to evaluate their performance. It was to recommend and help implement remedial action, whenever such action was called for. Moreover, a system of incentives and rewards for the managerial personnel, linking them to increased productivity or improved performance of enterprises, was recommended.

As to the appropriate criteria for the selection of the heads of corporations, the Five-Year Plan emphasised the qualities of dynamism, initiative and managerial experience as well as the knowledge of business ethos, methods and practices. The methods and procedures of a government department or ministry were seen by the Plan to be very different from those of a business corporation; the prevailing archaic methods of public administration in Bangladesh with their emphasis on undue centralisation, on the one hand, and status and precedents, on the other, were considered particularly inappropriate. The dislodgement of the civil servants from the positions of exclusive leadership in the corporations led to a reduction in their power and patronage, which was not taken kindly by them; it affected their attitude towards the appropriate system of relationships between the corporations and the ministry; it contributed to bottlenecks and indecision in working out a proper institutional framework for the management of the nationalised industries. In a situation where responsibilities and obligations of various agencies or functionaries were not clearly known, conflicts and tensions, with attendant delays, were frequent.

(f) The Performance of Nationalised Industries and the Role of Pricing Policy

The nationalised industries, on the whole, did not perform well in terms of generating surpluses and contributing to resources for development.[9] Out of the ten corporations, the biggest sector, the jute manufacturing industry, along with two others, i.e. paper and board, and tanneries, made consistent losses. Seven other corporations earned profits but at widely varying rates, only four having profits of more than 10% of the gross receipts from sales. Apart from deficiencies which resulted from the state of confusion, tensions and conflicts, surrounding the functions and responsibilities of the corporations *vis-à-vis* the Ministry of Industry, there were specific factors which contributed to their poor financial performance. Most prominent amongst these specific factors were disruptions in production because of physical damage caused by war, inadequate supply of electric power, transport bottlenecks, lack of entreprenurial or managerial ability, labour unrest and loss of traditional markets for a few export products. The level of capacity utilisation in most of the industries fell below the levels achieved before independence. This was partly due to shortages of raw materials, spares and components, most imported, and partly due to low levels of export demand. The need to service the burden of interest and amortisation payments on large, pre-independence debts led to additional borrowing, and a consequential increase in debt service payments and in deficits.

A major reason for losses and low profits was the pricing policy followed by the nationalised industries. This illustrated the difficulties encountered by them in the pursuit of the multiplicity of objectives enjoined by the government.[10] This was also a test case for the issue of autonomy versus government control in the management of the nationalised enterprises. The government was responsible for fixing the prices of output of most of the nationalised enterprises. The Cabinet, the Ministry of Industry and the Ministry of Trade were involved in various ways, either singly or jointly, in the fixation of prices for the public sector enterprises. Most of the products of the public sector industries consisted of essential consumption or intermediate goods. Under the Essential Commodities, Prices and Distribution Order, 1970, the Ministry of Trade fixed the prices of thirty-eight categories of commodities both domestically produced and imported. These products ranged from baby milk food, cement, edible oil, washing soap, cotton cloth, sugar, safety matches, to radio and transistors, electric lamps, artificial silk, synthetics, bicycles, toothpaste and electric fans, etc., many of which would not qualify as essential items in the context of prevailing poverty in Bangladesh. Although the prices of these commodities were fixed by the government, it had little or no control over the supply or distribution of the vast majority of them with the result that actual prices paid by the final users or consumers reflected their scarcity and were above the controlled price. For many products which were produced by both the private and public sector enterprises, frequently the prices fixed by the public sector enterprises were lower than those of their private counterparts; market prices corresponded to prices fixed by the private enterprises and as a consequence, the potential profits of the public enterprises were absorbed by the distributive channels.

Government controls over the pricing policies of nationalised enterprises limited their flexibility in matching revenues with costs. The product prices fixed by the government did not always cover their costs of production, and the consequent losses were covered by the creation of bank credit. Other measures to cover losses included *ex post* capitalisation, postponement of loan repayments, including the inadequate provision of depreciation funds. Escalation in raw material prices and other production costs reduced the profitability of many nationalised industries. A few nationalised enterprises which quickly adjusted their prices in response to rising costs did make profits, as for example, the steel industry and cotton textile industry. But there were others which adjusted prices to rising costs only after a time lag and suffered losses in the meantime.

In a few cases the government fixed the prices of both inputs and outputs; since in these cases the state enterprises could not buy in the cheapest nor sell in the dearest market, their profitability depended on the extent of technical and managerial efficiency. The price of sugar cane, for example, was fixed in order to keep the cost of the manufactured sugar low so as not to impose hardship on the consumers; at the same time if the price was not remunerative to the producers of sugar cane, then an adequate supply to the manufacturing industry could not be assured. A low price of sugar cane threatened to divert land to the competing crops or to encourage its smuggling to India or its diversion to the production of indigenous sugar. In the case of export industries the adjustment of prices to rising costs was constrained by the need to compete in the export market. The profitability of the export industry was also closely linked up with the exchange rate policy.

The appropriate pricing policy for the nationalised enterprises was debated at length. The opposition to the freedom of nationalised enterprises to fix prices in the light of their own evaluation of cost and demand conditions, was based partly on the ground that they enjoyed a monopolistic position in the domestic market, which was further buttressed by import controls ruling out effective competition from abroad. Therefore, it was feared that freedom to fix monopoly prices and earn high profits would only be at the expense of the consumers. This was not desirable for income distributional reasons. A restriction on their freedom to exploit the monopolistic market power was expected either to provide them with incentive or exert pressure on them to reduce costs and to improve efficiency. The possibility that individual enterprises in a particular industry could compete with each other in terms of variations of prices, differentiation of product and selling efforts such as advertisements, etc., was not a significant enough counterweight to the monopolistic position enjoyed by it. An appropriate system of material incentives and rewards for the manager and workers, which would link their economic prospects with the profitability of enterprises, could conceivably introduce an element of competition in improving efficiency.

It was, therefore, argued that in the absence of competitive market forces, such as free entry or free imports, and in the presence of considerable excess demand, the pricing policy of the nationalised enterprises should be formulated by an outside body, either an independent commission or the Ministry of Industry, on the basis of examination of costs of production and a desired rate of profit. However, this approach seemed to have several drawbacks. An outside body would not have access to an independent source of data on costs in all its details, which would be necessary for the fixing of prices; the public enterprises would have a built-in incentive to "doctor" or exaggerate costs, because this would enable them to have high and comfortable prices without any pressure to reduce costs. Determination of costs would face added complications in the case of joint products. Moreover, in order to fix appropriate prices the outside agency would have to make *ex ante* judgement about the expected level of efficiency of operations. Admittedly the prevailing level of efficiency was less than satisfactory. The various external factors such as the supply of imported raw materials, transport bottlenecks, power failures, etc., affected the efficiency and profitability of industrial enterprises. The incidence of these factors on their efficiency in the future was a matter of uncertainty and the outside body would be required to hazard a guess or make an *ex ante* judgement.

Price fixation by an independent agency was based on the assumption that with predetermined prices and with the requirement to earn a certain rate of profit, public enterprises would concentrate on economising costs. However, in practice, price fixation could not guarantee either profitability or efficiency; since it would be difficult to distinguish between the internal and external causes of losses or low profitability, such a system of price fixation could not use profit–loss account as an automatic check on efficiency. Losses suffered by public enterprises would have to be met somehow either by additional bank credit or by straightforward grants or subsidies from the government. Inefficiency and mismanagement might lead to the dismissal of managers only under exceptional circumstances. The ultimate sanction of bankruptcy could not be used against public enterprises. Moreover, the process of price fixation by an outside agency on the basis of examination of cost data was very time consuming; by the time prices were fixed, the cost data were likely to be out of date. This was particularly true in a period of inflation with a rising trend in the general level of prices and costs. Delay in the examination of costs and in the fixation of prices would result in further losses for the nationalised enterprises.

The planners in Bangladesh recommended a middle course between complete freedom for public enterprises to fix their own prices and fixation of prices by an independent agency. Within certain predetermined limits, the enterprises should have the freedom to fix and change prices, which should be subject to *ex post* scrutiny by such an agency. The limit was set at 15% over costs of production. If the independent agency which would monitor prices and costs on a regular basis found the price of a commodity on an *ex post* examination to have been based on incorrect or inappropriate costs, it would recommend a review. The enterprises could revise or refix prices on their own in response to changes in costs, so long as the revised prices were within the limit of 10% of the original prices; such price changes would be subject to a subsequent review or *ex post* examination. In case a particular enterprise was in need of a subsidy in order to compete in the export market or to meet temporary financial difficulties, the case was to be referred to the independent agency. If excess demand and scarcity resulted in a wide gap between the ex-factory price, determined along the above lines, and the ruling market price, the ex-factory price might be so fixed as to mop up the scarcity margins. The excess profits could accrue to the exchequer and could augment the general pool of resources available for development. This policy was, however, not without dangers. It was based upon the recognition that there was no effective *ex ante* safeguard against overcharging or high cost and inefficiency on the part of public enterprises; the only safeguard was that of *ex post* evaluation. Even after *ex post* evaluation, it would be difficult to be conclusive about the appropriateness or otherwise of the prices fixed by public enterprises. Evaluation of many of the factors underlying the fixation of prices would be a matter of judgement. Moreover, the restriction of the revision of prices to the upper limit of 10% of the original prices ran the risk of inflicting losses on public enterprises, in an inflationary situation, in the intervening period when the adequacy or otherwise of such revisions was being scrutinised.

There was no easy way out of the dilemma involved in seeking, on the one hand, to reduce inefficiency and, on the other hand, to prevent losses arising from the inflexibility of prices. The best that could be accomplished under the circumstances would be to pursue one clear-cut objective and to require every public enterprise to earn a profit. The public enterprises should have the freedom to charge prices in order to make a profit and to generate a surplus for investment. Losses would lead to deficit financing or the creation of additional bank credit; it was likely, furthermore, to lead to the neglect of repair and maintenance of capital and depletion of stocks. The requirement to earn profits was expected to put pressure on public enterprises to respond to changes in costs and demand, either by changing prices or by improving efficiency. Under these circumstances if public enterprises suffered losses, they should be called upon to explain the reasons thereof and to open themselves to detailed investigations at the hands of the independent monitoring agency. It might be possible to devise supplementary or additional checks to prevent the use of high prices for making excessive profits or for covering up inefficiency. For one thing, the schedule of selling prices, including all price changes, should be widely publicised. Moreover, if an industry raised prices too frequently, it should be called upon to explain the causes of such action in public. Keeping the prices under constant public view and discussion was a way of keeping pressure on the managers of public enterprises. Furthermore, the banking system could act as a watchdog on their activities by enforcing discipline in financial operations; this required that the banking system, which was nationalised in Bangladesh, was to be enjoined to deal with the public enterprises on strictly commercial principles.[11]

Moreover, even if public enterprises earned a profit, they might be required to satisfy additional criteria of efficiency, such as a favourable comparison with international prices; the trend in relative prices, prices of domestic enterprises in relation to world prices, was expected to be downward, since "infant industry" handicaps were to be overcome over time. A scrutiny of inter-firm or inter-enterprise price differences was a good check on the reasons for inefficiency or losses in the particular enterprises in the same industry and, therefore, on the general level of performance of the industry. Indices such as output per man hour or per unit of capital could be compared with those in other enterprises or industries under similar conditions, both inside and outside the country. The *ex post* investigations of indices of efficiency along the lines mentioned above could appropriately be carried out by an independent agency on a continuous basis.

(g) Incomes Policy

The political leadership in Bangladesh made frequent references in its policy pronouncements on the future economic system to the need for the introduction of a just and egalitarian society. The nationalisation measures, including the ceiling on private investment, need to be viewed in this overall perspective.

In this context the following references to the Finance Minister's budget speech in 1972 might be appropriate:

A vital aspect of a socialist society free from exploitation is rapid economic growth and just and equitable distribution of the fruits of such growth. During a phase of rehabilitation and reconstruction, it is expected that emphasis should be accorded to economic growth. The relation between growth and equitable distribution of its fruits cannot, however, be denied. Policies adopted in order to fulfil the above objectives include the nationalisation of industrial and other enterprises, cuts in salaries of high salaried persons both in the government and private employment, in tax exemptions and tax adjustment measures. To remove inequalities of income and to rationalise the pay structure, the government is setting up a Pay Commission which shall start its work very shortly.[12]

In 1973 the Pay Commission recommended measures to reduce wide differences in the pay and salary scales in the public sector, both in government departments and in public enterprises, mainly by compressing the higher income scales.[13] The Pay Commission argued the case for a minimum living wage which was necessary to meet the cost of a minimum basket of basic goods and services. It determined the minimum pay scale on the basis of existing prices and recommended the supply of essential goods at subsidised prices, prevailing at the time. It recommended ten grades of pay, with a maximum salary ceiling of Tk.2,000 per month and fringe benefits of Tk.300 per month, and the minimum pay of about Tk.200 and fringe benefits of about Tk.80 per month. The highest income was no more than ten times the lowest. The recommended salary scale involved an absolute decline, compared to the pre-independence period, in incomes of the public employees in the higher grades. In the months immediately following independence, i.e. in early 1972, the ceiling on incomes was put at Tk.1,000; it was subsequently raised to Tk.2,000 in view of rise in prices. The ceiling on incomes of the employees in the public sector were justified on the grounds that "as long as unequal opportunity for development of individuals' innate capability is continued, it will be unfair and socially explosive to allow very wide differences in incomes, including that in salaries of public employees".[14] Since the lowest pay could not be substantially increased due to the paucity of resources, the highest pay was to be brought down within reasonable and tolerable limits. The Pay Commission recognised that the highly qualified technical and professional skills could not be adequately rewarded within the recommended ceilings, especially in view of the attraction of the international job opportunities; therefore, compensation should be provided in terms of high social status or high authority in the hierarchy of government, moreover, non-material incentives such as patriotism and spirit of social services should be encouraged.

The imposition of a ceiling on salary income was not the only way of achieving a reduction in inequality in such incomes. It could be brought about by a policy of progressive taxation as well. Moreover, the tax mechanism could extend to all incomes, i.e. salaries, rents or profit incomes, and cover both the public and private sectors. Unlike the ceiling on salary incomes, it could be non-discriminatory in its treatment of all categories of incomes. It was further decided by the government that salary incomes in the public sectors would be tax free. The inequalities in non-salary incomes, which would be subject to tax, were to be reduced by the instrument of progressive taxation. But politically and from the point of view of popular appeal, the income equalising effects of progressive taxation were not readily seen or were too sophisticated. Low differentials in incomes received, even though they related to tax-free incomes which were readily visible to the public eye, had a greater political advantage or effect than high differentials in pre-tax incomes, which would be reduced by progressive taxation.

However, ceilings on salaries, in spite of the exemption from income tax, became a matter of complaint by the public sector employees; it was argued that for psychological reasons as well as from the point of view of incentives, high pre-tax income was preferred to tax-free low income, even though the former would be brought down to the level of the latter after taxation. The tax experts contended that the exemption of salary income from taxation might provide scope for possible evasion in the case of salaried persons who had non-salary incomes.

Irrespective of the methods by which salary incomes of the public sector employees were squeezed, whether by direct ceilings on income or by progressive taxation, the incomes in the private sector, including profit incomes earned in commerce and industry and incomes of the self-employed, were not directly affected. Especially high incomes were being earned in the distributive trade and related private activities, owing to prevailing high scarcity. These were also the categories of incomes on which it was relatively easy to evade taxes. This created disaffection among the fixed-income earners in the high-income brackets in the public sector.

The members of the civil service, including the military, were adversely affected in terms of income and material benefits which they had traditionally enjoyed. A decline in the level of their material comforts, both absolutely and in relation to other public sector employees, caused tension and dissatisfaction. Questions were raised as to whether a decline in real income did not contribute to an increase in the incidence of corruption or whether the latter was due to increased opportunities for graft. Both the opportunities and magnitude of personal gains to be secured through the abuse of power in a regime of acute scarcity and wide ranging discretionary controls, were in all probability higher than before. While there were those who were likely to exploit such opportunities, no matter what was their level of income, undoubtedly there were others at the margin whose susceptibility to corruption or ability to resist the temptation for corruption, was affected by a decline in their relative or absolute income. The demonstration of high living and consumption standards on the part of those who had access to high incomes in the private sector or who were corrupt, heightened the public awareness and complaints about the inequities resulting from prevailing policy. This created a credibility gap with respect to the socialistic and egalitarian professions of the government.

The managers of state enterprises were faced with a particular dilemma; they had direct access to and control over considerable financial resources; their opportunities for securing private gains through the abuse of power which they exercised were considerably greater than in the case of civil servants in the government departments or ministries. Many of them were previously employed in private industry or trade and were accustomed to much higher levels of living than the general run of the civil servants. They suffered a substantial decline in real income. At the same time any slackening of incentives and initiative on their part caused by a fall in real income, could have serious and immediate consequences for the efficiency and productivity of state enterprises. They had no ideological commitment towards socialism nor were they expected to behave like the cadres of a socialist party or a socialist state. As professional managers they were used to responding to material incentives; their loss of material comforts and incentives was not compensated for either by a new-found ideological motivation or by an increased status, greater independence, higher responsibility and authority.[15] Moreover, the high rate of inflation in the first three years of independence aggravated the lack of correspondence between the real income of the public sector enterprises, on the one hand, and their status responsibility and authority, on the other.

In so far as the relative position of the workers was concerned, even though trade union pressure increased money wages, this was not accompanied by an increase in productivity. Inflation and industrial unrest, resulting in a loss of production and in higher money wages, fed on each other. As described earlier, in the period immediately following liberation, several factors contributed to an increase in industrial unrest. During the last days of the Pakistani rule, strikes and threats of violence by industrial workers were used as a weapon of non-co-operation with and struggle against the regime and its supporting Pakistani industrialists. Their participation in the struggle for liberation also heightened their expectations.[16] There were a large number of labour unions competing for the allegiance of the workers not always through peaceful means of persuasion but also through the threat of a resort to physical violence. They had become used to violence during the war, when they also had acquired arms.

The labour leaders acknowledged that the dislocation caused by labour indiscipline had serious consequences for the economy. They however, argued that the workers could not be expected to exercise constraint in their demands for higher material benefits, if sacrifices were not equitably shared by all sections of the community. In the past the surplus generated in the productive sectors, often at the expense of constant or declining real wages, was partly, if not largely, spent on conspicuous consumption of the industrial and commercial entrepreneurs as well as of the top members of the bureaucracy. There was a strong opposition to any repetition of the past experience; they wanted any restraint exercised by them to be reflected in investible surplus, which in turn should be seen to be used to increase the supply of basic goods and to expand employment opportunities.[17]

The expectations of workers, raised in the wake of independence, were not to be fulfilled. The workers did win several wage increases, a major one in 1972 and again in 1974, and in between they obtained various ad hoc increases in fringe benefits, sometimes by negotiation and sometimes by threat of violence. At the same time, however, they found their gains in money wages eaten up by inflation. In the three years from 1971/72 to 1974/75, real wages of the industrial workers in the nationalised enterprises declined by about one half, roughly in line with what happened to real wages of the agricultural workers.[18]

The proper role of trade unions in the public enterprises in Bangladesh remained an unresolved issue. The labour policy as announced in 1972 stated that in the public enterprises there was no scope for more than one union or for collective bargaining. The trade unions were merely to serve as welfare organisations rather than as instruments of bargaining or of determination of wages and conditions of work. Under these circumstances and particularly if workers were to participate in the management of state enterprises, strike as a weapon of bargaining was to be considered unnecessary and illegal. The general level of wages and other benefits was to be determined at the national level by negotiation between the representatives of management and workers of all the nationalised enterprises as a whole. This policy was opposed by the workers' organisations. As a result, its implementation was postponed until an agreement or a consensus was reached between the government and the workers' representatives.[19]

The case against the multiplicity of labour unions and their role as instruments of collective bargaining was built on two counts. Firstly, it was argued that in the public enterprises there could be no conflict or dichotomy between capital and labour in the same way as it existed in the private enterprises. The public enterprises were conceived to be owned socially; the workers no less than the management had a stake in their increased productivity and profitability. The level and structure of industrial wages was an integral part of general incomes policy, which would spell out the principles of determination as well as the relative income shares of different economic groups in society. This would discourage an attempt on the part of any particular economic group to gain advantages for themselves without considering the consequences of such action on other sectors and groups of the economy. Secondly, in a system of workers' participation in management, a multiplicity of unions engaged in winning support among the workers would render impossible the role of the labour representatives on the management board. Rival unions finding themselves in the minority would either force the workers involved in management to advocate or support the escalation of workers' demands or would make such representatives appear to the workers as tools of the management. With only one union and its representatives sitting on the management board, they should be able to examine the question of the workers' relative share in the income of an enterprise in light of the overall requirements of reserves and reinvestment, the need for contribution to the general revenues of the state as well as the provision of material incentives for management.

The dilemma faced by Bangladesh in defining an appropriate role for the trade unions emerged from the characteristics of the political and economic situation in Bangladesh. In the first place, even though Bangladesh proclaimed itself as a socialist state, it was as yet a mixed economy; there was no unanimity or certainty regarding its evolution towards a socialist state with none or very little private property. As it was, private property was quite dominant outside of the large-scale industry, foreign trade and financial institutions. Pressures were constantly exercised against the socialist policies; the working-class movement was very far from being in control of the apparatus of the ruling party and the state machinery to a degree even distantly comparable with what had been achieved in the socialist societies. In general, it did not feel confident about the prospects of socialism in the country. Therefore, it was not possible to transfer to Bangladesh the principles of labour organisations obtaining in such socialist societies. Moreover, private and public sector enterprises co-existed in trade and industry and the workers' organisations were interlinked between the private and public enterprises. While the trade unions with their rights of collective bargaining were to function in the private sector, they could not behave differently in the public sector without creating splits in their own ranks. Bangladesh also proclaimed herself as a multi-party democracy. The trade unions were closely linked with the political parties; the mobilisation of industrial workers was a very important component of the strategy of the political parties for winning support among the masses. In the context of the competing political parties, it was unrealistic to advocate the elimination of the competing trade unions.[20]

In the light of the above circumstances the workers did not feel very enthusiastic regarding the proposal about their participation in management. In the first place, they had to surrender their right of collective bargaining and the multiple unions. Moreover, in order to sit on the management board, their representatives had to be employees of the same enterprise. Therefore "outsiders" or the professional labour leaders, who had traditionally provided their leadership could not represent their interests in the management board. This was unacceptable to the unions, since they maintained that within each and every enterprise the workers would not be able to elect members who had enough knowledge, expertise or experience to protect their interests in the management board, especially since the workers' representatives would be in a minority.

A variety of compromise solutions were discussed from time to time. It was proposed, for instance, that while the workers would have the right to freedom of association and to the multiplicity of unions, they should be required to determine by a majority vote which of the unions should have the status of "exclusive, recognised union" for a stated period. This union would then act as their representative for the purposes of collective bargaining and participation in management. This would have required an agreement by the labour leaders to contain their intra-union rivalries. In the second place, collective bargaining at the level of plant or industry in the public sector could be allowed with respect to such matters as general conditions of work, labour training, safety of work and welfare measures, etc. Even in this restricted sphere collective negotiation and bargaining should not end in strikes; if negotiation failed, it should be submitted to compulsory arbitration. The participation by workers in management, therefore, was intended to involve them in all aspects of management and operation of enterprises and not necessarily to protect their sectional interests. The determination of wages and other material benefits of workers was related to the incomes policy for the economy as a whole, including the socio-economic development strategy, in the discussion and determination of which the workers equally with other sections of the economy should participate at the national level.[21]

None of these proposals reached the stage of agreement between the government and the workers' organisations. In the meantime, a multiplicity of unions continued to function in the industrial sector; a variety of unions, based on regional, political and personal affiliations, vied with each other for the loyalty of the workers, and were engaged in escalating demands for wages and other material benefits. In the pre-nationalisation days, private entrepreneurs could effectively "divide and rule" the several unions and thus managed on the whole to keep industrial peace; by providing material rewards to the rival union leaders they bought peace or broke up workers' movements. The nationalised enterprises could not effectively resort to these expedients in order to buy industrial peace. What was needed was a national consensus through political negotiation between the labour unions and the political leadership to delineate the broad outlines of incomes policy. In Bangladesh such a consensus could not be achieved more than in other developing countries; at the same time, the general economic situation deteriorated and income distribution worsened.

Thus the economic fortunes of the fixed-income earners, including civil servants and managers of public enterprises as well as industrial workers and the rural poor, all declined, while the economic position of traders, private industrialists, and surplus farmers improved. At the same time, there was a marked increase in conspicuous consumption on the part of a limited number of high-income earners in the big cities. The promise for a broad consensus based on democracy and socialism remained unfulfilled.

(h) A Critique of the Industrial Investment Policy

Right from the beginning, the scope and limits of private versus public enterprise came in for criticism from various quarters. The industrial policy was considered the touchstone or the symbol of the government's attitude towards the future of the economic system. The debate really never rose, either within the party or in the country at large, to the level of a debate of the socialist goals of the country. The criticisms against the investment policy originated from different components of the ideological spectrum, from the right as well as from the left. Because the debate prior to the announcement of the policy in 1972 was not conducted in public but in camera in the Cabinet, and also because it was not very exhaustive or conclusive, these points and counterpoints kept cropping up through the following years.

There were those outside the ruling party who generally were in favour of an expanded public sector in Bangladesh, including a few who professed their faith in the socialist future for Bangladesh. But they were divided in their attitude towards the nationalisation measures in 1972. To a few of them the expansion of the public sector appeared to be too hasty; in the first instance this should have been confined to the abandoned enterprises. The nationalisation of enterprises owned by Bangladesh nationals could have been postponed to a later stage. Others argued that the essential and necessary condition for the achievement of socialism was not so much the extent and nature of nationalisation as the existence of an ideologically-oriented party with dedicated cadres, to inculcate attitudes and incentives relevant to a socialist society. Without the working classes being in control of the state apparatus, and without a clear and strong ideological motivation of the political leadership, it was argued that a policy of nationalisation could contribute to the benefit not of the masses but of the private interests which were dominant in the rest of the economy and consequently in the political system.

The criticism from the left was, therefore, that the middle class, dominant in the economic and political system of Bangladesh, would reap the benefits of nationalisation in the same way that it benefited from other economic policies and indeed from the working of the economic system as a whole. The state enterprises were not likely to be an exception in this respect. They did expand the employment opportunities for the middle class in white-collared jobs; the output and pricing policy of state enterprises did favour the urban consumers and affluent farmers in the same way that the fiscal, monetary and credit policies did.[22]

There were indeed certain instances of policy in the early years in Bangladesh which lent credence to this line of attack on public enterprise. These policies were not inherent in the act of nationalisation but were mainly the results of political ineptitude, administrative inefficiency and misuse of political power—all combined together in the highly abnormal circumstances prevailing immediately after independence. The appointment in early 1972 of a number of administrators for the abandoned industrial units, who had no previous knowledge or experience of industrial management was one such instance. The result was not only mismanagement but also personal aggrandisement on the part of many of them; this was facilitated by the fact that no inventory or accounting of assets and liabilities was made before the enterprises were handed over to the administrators. The fact that within a matter of days after independence, the government was faced with the task of appointing hundreds of administrators for the abandoned industries, did not help careful selection. Also the administrators in many cases appointed marketing and sales agents for the enterprises under their management on personal or political considerations, which further aggravated the opportunities for personal gain or corruption. Even after the establishment of the corporations, a few of the initial administrators continued and a number of ex-owners were appointed as managers; being unsure of their long-term future, they sought to maximise their personal profits rather than the long-run viability of the industrial enterprises under their management. The appointment of ex-owners as managers of their own enterprises was probably a mistake; they should have been appointed in enterprises other than their own and, better still, in the sector corporations which were in charge of the overall regulation of state enterprises.

The architects of the industrial investment policy were clearly in an unenviable position, subject to attack, as they were, from all sides of the political spectrum; they themselves had only half-hearted convictions. To them the chief justification for the nationalisation measures and for a ceiling on private investment was to create and strengthen the objective conditions which, if they did not usher in a regime of socialism, would at least thwart the growth of a large capitalist class.

The two most important criticisms, advanced principally by those who opposed on ideological grounds the expansion in the role of public enterprise, were that it discouraged savings and capital accumulation and that it allowed such entrepreneurial and managerial talent as was available in the private sector to go to waste. Whether the ceiling on private industrial investment acted as a disincentive against private savings was not an easy question to answer. It was arguable that such investment opportunities as existed within the limit of Tk.2.5 million, combined with the investments in the smaller, abandoned industrial units which were available for purchase by the private investors, were large enough to absorb all private savings seeking such investment outlets. This was a factual question on which evidence was necessarily inconclusive. In fact, there was a great deal of debate as to whether the industrial units with assets up to Tk.2.5 million were economically viable, and, if they were, whether there were many such enterprises. However, during 1973/74 and 1974/75, the government licensed one hundred and seventy-two enterprises with fixed assets per unit up to Tk.2.5 million; about half of the total investment was sanctioned in units of assets below Tk.1 million.[23]

Similarly, whether private investment was discouraged by the ceiling was unclear. As many as seven hundred applications for loans were submitted to the public financing institutions during 1973/74 for investments within the ceiling. This was consistent with the hypothesis that the investment policy did not discourage small entrepreneurs willing to invest their savings. However, the counter arguments that many of these applications were spurious, that they did not really represent serious investment intentions, that in any case many of them either did not fulfil all the requirements of eligibility for loans or underestimated costs, could be met or verified only after the examination of the loan applications was completed by the financing institutions. Therefore, the controversy could not be resolved until after some experience had been gained in the implementation of the policy.[24]

In so far as the opportunities of private investment were concerned, they were not confined to the industrial sector but included agricultural and related activities as well. Already opportunities for investment in the agricultural sector such as irrigation pumps through co-operatives and shallow tubewells under private ownership were expanded, in some cases, with credit facilities from the government. Furthermore, private savings could in part be placed in financial assets; with this end in view after some hesitation and time lag, interest rates were substantially increased. However, with a high rate of inflation, the real interest rates remained very low. The need for continuous improvement and expansion of financial institutions and instruments for the mobilisation of private savings remained paramount.

At the same time, however, there were many instances of either lapses in implementation of policies or sheer inefficiency and indecision which contributed to a feeling of uncertainty among the private investors. One was the vacillation in policy regarding the savings certificates issued by the Pakistani government, including those purchased during the war of liberation by the private savers in Bangladesh. Whether all or most of these certificates would be honoured by the government of Bangladesh was not decided for a long time. Even after a positive decision was made to honour all certificates, except those purchased during the war of liberation, there was a delay in redeeming those which were eligible and due for redemption. There was also a delay in the payment of compensation which was promised to the owners of enterprises which were nationalised in 1972; this added to the feeling of uncertainty among the private investors regarding the intentions of the government about the future of private enterprise. It was in this general atmosphere of uncertainty that the critics of the investment policy came out strongly against the moratorium on nationalisation for a period of ten years which was announced by the government. It was argued, paradoxically enough, that the moratorium against nationalisation acted as a disincentive to private investment; the intention was, however, the reverse, i.e. to assure the private investors that in no case would their investments be nationalised within the ten-year period. The time period was considered long enough for most private investments to be recouped with reasonable profit so that they would not stand to lose, even if their investments were nationalised after ten years. Moreover, there was no implication that even after ten years the existing private investments would be nationalised; it was an open question and the government had the freedom to exercise its option to nationalise, if it at all wanted to do so. The critics advocated either a considerable extension of the guarantee period or its elimination on the ground that a ten-year moratorium instead of assuring the private investors created an apprehension that all private investments would necessarily be nationalised after ten years. One could, however, argue that any such guarantee which was not a constitutional guarantee but only a policy statement by an incumbent government could be reversed by a successor government. Therefore, the guarantee could be ineffective and was not sufficient by itself to provide firm assurance against nationalisation in the future. But it was one thing to say that the guarantee was insufficient and quite another thing to suggest, as the critics suggested, that the guarantee itself acted as a disincentive. It was only the prevailing atmosphere of uncertainty and debate between the contending forces for and against private enterprise which could possibly account for such a specific interpretation of the declaration of moratorium on nationalisation.

(i) Pressures for Revision of Industrial Policy

Starting from 1973/74 there was increasing pressure for an upward revision of the ceiling and for allowing a larger role of foreign investment, especially its collaboration with domestic private enterprise. It is worth exploring reasons for and forces behind such pressure.

By 1974 there were a number of factors which had contributed to an accumulation of surplus funds in private hands. For one thing, high profits were earned in domestic and import trading activities, including illegal trade such as trade in contraband goods and in smuggling jute and other exportables across the border. Since these transactions were illegal, the risk premium was high and hence profits, once realised, were high. In addition, many residential buildings and trading or commercial enterprises, abandoned by Pakistanis, were illegally occupied by private persons, in some cases under the cover of political protection. The "caretakers" of such commercial enterprises, hastily appointed by the government in 1972 immediately after independence, made large fortunes through the undeclared sale of assets. Moreover, there were gains to be obtained from rental or sales proceeds of the abandoned houses which were illegally occupied by private persons.

Those who had accumulated financial resources were pressing

the government to commit itself to a more substantial and permanent role for private enterprise in the economy of Bangladesh. An upward revision of the ceiling on investment in individual enterprise as well as an enlargement of the private share in total industrial investment, was looked upon as a confirmation of the government's positive attitude towards encouraging a thriving private sector in the future. In this context, whether existing opportunities for private investment in the industrial sector were adequate or not to absorb available investible funds, was not a major issue. What was crucial was that the government reverted, gradually, if not immediately, to the pre-liberation policy of "sponsored capitalism" where the state would provide concessional financial and fiscal facilities to allow the private investors to invest, to reap large gains, to gain initial experience, if necessary with government subsidies, and then to grow in size and power. They were eager to ensure that private enterprises were not confined merely to the small-scale industries—a small island in an industrial sector owned largely by the government, which remained publicly committed to the establishment of socialism. There was pressure not only for raising the ceiling on investment but also for declaring a moratorium on any action against ill-gotten or "black" money, i.e. money earned in illegal transactions such as smuggling and black marketing and avoidance of taxation. All hidden wealth was to be exempted from penalties for past violations of foreign exchange and tax laws, provided these funds were invested in industrial ventures in the priority areas prescribed by the government.

It was not, however, difficult on the part of the supporters of the existing policy to produce counter arguments against "legalising" black money. In the first instance, it would have accorded legitimacy to the violation of government laws and regulations; it might indeed have encouraged such activities in the future. To exempt or reward those who made fortunes by a violation of laws in the context of a general state of deprivation in the economy was against the prevailing ethics and expectations of a society; a political party which was alleged to be committed to an egalitarian society could not have easily done it. To fail to enforce laws and regulations or even to participate clandestinely in their violation for private gains, was different from according public and social approval to illegal and anti-social acts. Moreover, the economic advantages of a policy of legitimising black money were neither clear nor overwhelming. Whatever black money was made in the early years, especially in 1972 and 1973, would already have been spent, in all likelihood, on high living, real estate and precious metals or transferred abroad, as witnessed by the heavily depreciated rate for Bangladesh currency in the black market for foreign exchange. If private assets were already held in cash hoards or idle bank deposits, their activisation was merely inflationary; a better course would be to immobilise such funds by the offer of attractive interest rates on financial assets or, as was done in early 1975, to cancel them at least partly through demonetisation of high denomination currency notes. The best way of dealing with black money, in this view, was to deal with the sources of generation of "black" money by reducing the gains from illegal transactions. Moreover, there was the risk that an upward revision of the ceiling might increase, not reduce, the scope for illegal gains. For instance, by inflating the cost of land and buildings and by resorting to bank borrowing against the security of their inflated value, the investors could reduce their actual contribution to and stake in total investment. Similarly, a higher ceiling could provide a greater scope for over-invoicing the cost of imported capital goods and for securing easy access, at the official rate of exchange, to an additional amount of foreign exchange which could be transferred abroad or sold in the black market.

In sum, the supporters of the ceiling argued that a revision of the ceiling was neither necessary nor sufficient for increasing the flow of private investment and certainly not in 1974. There were a number of factors unrelated to the level of ceiling which might have adversely affected the climate for private investment. For one thing, there was continued militancy among the industrial workers. To make matters worse, there was no great improvement in law and order since 1972. There was uncertainty about the future political system; there were apprehensions that tensions between the different political parties as well as between the rival sections of the ruling party might erupt into open violence and thus threaten economic and political stability. In such circumstances private investors were shy and were likely to seek short-gestation or quick-yielding investments or to transfer capital abroad. The scope for private investment was further hampered by the shortage of foreign exchange. The ongoing development projects, mostly in the public sector, pre-empted a large part of the available foreign exchange resources for financing investment. Also, there was a pressing need for foreign exchange for imports of raw materials and spares to provide a fuller utilisation of the already installed capacity. A revision of the investment ceiling by itself could not have removed these constraints on private investment.

The pressure for raising the ceiling on private investment exerted by the new rich within Bangladesh was strongly and consistently supported by foreign aid agencies which advocated a larger role for private enterprise in the industrial development of Bangladesh than was provided for in the Five-Year Plan; thus the World Bank recommended raising the ceiling to Tk.10 million or removing it altogether.[25] At the same time, foreign investors were pressing for opportunities of collaboration with the domestic private investors. Also, within the government there was a strong group among the members of the bureaucracy and the civil service in favour of an upward revision of the investment ceiling. To many of them a private enterprise system, subject to controls and regulations which gave them power and authority, was a preferred alternative to public ownership.

It is clear from the foregoing analysis that though the industrial policy statement of 1972 had laid down the policy objectives with respect to the evolution of the desired economic system of Bangladesh, the issue was not closed. In the rank and file of the ruling party and amongst its leaders, as already noted, there was no unanimity and no firm commitment to the extension of public ownership or the ultimate goal of socialism. On the one hand, the conservative elements in the party consisting of the trading classes, the aspiring industrialists and the ex-owners of the nationalised enterprises were all pressing for a revision of the policy. On the other hand, the radical wing of the party, consisting of the militant factions of the students and trade unions, were pulling in the direction of a further extension of public ownership. The Prime Minister stood in the middle of the contending forces. He responded carefully to the conflicting pressures—balancing off one step in one direction by another in the opposite direction. He sought to respond to radical pressures without alienating or provoking an extreme reaction from the conservative groups. The decision to nationalise in March 1972 was his considered response to the radical forces within and outside his party. It was based upon his judgement that the conservative forces on the right at that time were relatively weak and had no alternative leadership to fall back upon. However, he was unwilling to advance farther along the road advocated by the radical forces. In fact, a concession to the conservative forces appeared appropriate at that stage at least to stem the rising tide of criticism against the government's economic performance.

The criticism against the government's investment policy gained momentum as the economic position in the country worsened. There was a desperate search for policy measures which could in any way stimulate production and employment. The poor financial performance of public enterprises was often cited as an instance of how the government's predilection towards public ownership—a vital instance of misguided economic policy—had contributed to economic distress. The policies of the government, including the constraints placed upon the operations of public enterprises in terms of output, employment and pricing policies, as well as the external and internal economic factors operating on the Bangladesh economy, which contributed to their poor performance have already been discussed. Whatever the complicated set of factors leading to losses in the nationalised industries, the image set in the public mind was one of inefficiency which not only caused a heavy drain on the public exchequer but also added to the overall inflationary pressure. This contributed to a loss of confidence in the role of the public sector as a means of generating growth and efficiency. At the same time the political leadership had started to have doubts as to whether its policy of a fixed ceiling on private investment was not in fact a cause of stagnation in the economy. If a revision of the ceiling was likely to stimulate a spurt in economic activity and an expansion of output and employment, this should be allowed. At least it would provide a test as to whether this measure would work in the way the critics insisted that it would. Eventually the industrial investment policy was revised in July 1974.

(j) Revision of Industrial Policy

The new policy reserved eighteen sectors in the public sector, the rest were thrown open to the private sector. Public utilities, such as telegraph, telephone, cables and wireless apparatus, generation and distribution of electricity, air transport and shipping, atomic energy, arms and aummunition were reserved for the public sector.[26] But inland shipping and road transport were thrown open to both private and public enterprise. In the field of industry proper, only eleven industries were reserved for the public sector: ship-building and heavy engineering, heavy electrical, minerals, oil and gas, petro-chemicals (including only fertilizers, PVC and ethylene and synthetic fibres), heavy and basic chemicals and pharmaceuticals, and iron and steel. Many of these industries were in any case beyond the capacity of private enterprise, because of large capital requirements, or they were unattractive to private enterprise owing to long gestation lag and low profitability. The only industries which were not exclusively in the public sector in the pre-independence days and which were now reserved for the public sector were the jute and cotton textiles, sugar and cement industries.

At the same time the ceiling for private investment was raised very substantially to Tk.30 million. Moreover, private foreign investors were allowed to enter into partnership with domestic private investors, mainly but not exclusively in projects where 'technological and managerial gaps existed'. Finally, the moratorium on nationalisation was extended from ten to fifteen years and fair and equitable compensation was guaranteed in the event of subsequent nationalisation.[27]

Between the middle of 1974 and November 1975 a number of

important political changes took place. In January 1975 a presidential system was introduced based on a single party which also included elements of the leftist or radical parties. It abolished parliamentary democracy with its attendant political freedom—one of the pillars of the 1972 constitution of Bangladesh. The nature, causes and significance of the political changes introduced in early 1975 had been variously interpreted. In 1973 and 1975 a number of dissident radical members of the ruling party, i.e. especially the radical militants amongst the students and labour unions, had left the Awami League (the ruling party) and formed a new party, Jatiya Samajtantri Dal (National Socialist Party) with a radical socialist programme. The prevailing economic stagnation, accentuated by rapid inflation and marked by a severe famine in the third quarter of 1974, seemed to increase the support of the radical party among the masses.

Among the observers of the contemporary scene, there were many who believed that this was a reaction by the leader of the ruling party to the leftist challenge as well as to the increasing unpopularity of the ruling party. However, it was not quite clear whether the subsequent shift in policies would eventually have been carried towards the right or the left. Before the scenario could unfold, there was a coup in August 1975, when the President (formerly the Prime Minister) was killed and a military government was established. In subsequent months there were two more military coups. The military government which was established after the coup in November 1975 opted for a much greater role for private enterprise in all spheres of economic activity, including both trade and industry.

The industrial investment policy was again revised. Industries previously reserved for the public sector were now thrown open to joint participation between public and private enterprise, yet with the public corporations retaining majority shareholdings. Moreover, "on considerations of economy, technology and management, management contract might be allowed to (domestic) private investors, wherever possible".[28] The provision regarding the moratorium on nationalisation was abolished with a view to removing remaining misgivings on this account in the minds of the investors. Instead it was reiterated that if any industry was ever nationalised, compensation would be paid on a fair and equitable basis. In addition, a few of the policies adopted in the pre-independence days to promote private capitalism were revived. These were mainly in the area of financial and credit policies. Generous financial assistance was provided not only in the form of loans but also in the form of equity. The provision of equity capital by the state to supplement private equity in a private enterprise was an important component of industrial policy in the pre-independence days as mentioned earlier. Furthermore, the public sector corporations which had acquired some of the abandoned but smaller industrial units were encouraged to sell them back to the private sector.[29] Also, "with a view to bringing the unutilised and idle funds into the productive stream and for mobilising investment funds for investment or purchase of disinvested units, it was promised that those who would disclose such funds would be allowed to retain the same without any questioning by the government provided they were used for new investments and purchase of disinvested units".[30] Subsequently, decisions were taken to disinvest or transfer to private hands the abandoned enterprises wherever found suitable, without limit as to the size or scale of operations.

In a résumé or a post-mortem of the "battle" over the question of a ceiling on private investment, which became a symbol of a much bigger debate on the pattern of the future economic system of Bangladesh, it might be worth while to summarise the nature and the objectives of the contending forces. Among those who supported the maintenance of the ceiling on private investment fixed in 1972 were those who genuinely believed that there were enough investment opportunities for the time being for the small investors within the limits of investment set by the government. Even by 1974, though inflation might have whittled down somewhat the real value of the monetary ceiling, given the shortage of foreign exchange resources and the pressing, overriding requirement of the ongoing public investment projects, private investment opportunities within the limits of the ceiling were considered adequate; more than this could not have been implemented with the existing foreign exchange resources. This group, furthermore, was in favour of raising the ceiling on investment over time, as and when the rate of inflation substantially reduced the real value of the monetary ceiling. They were willing to examine whether in fact there were a large number of small savers who were willing but were unable to obtain opportunities for investment—taking both rural and industrial investments together—and take appropriate measures for mobilising savings for investment. They were not, however, convinced that time was ripe in 1974 for such a re-examination. They wanted to promote private investment in the small-scale industrial sector but were unwilling to encourage private investment to outreach and grow into large industrial fortunes resulting in concentration of economic power and high income inequalities. Moreover, in their view the political and social environment, including a stable labour force and improvement in law and order in the country, was critical in determining the investment climate; an investment ceiling in its view was only one of a large number of factors which were likely to affect the quantum of private industrial investment activity. They constituted what might be called the middle group with a large following among the political leadership as well as among the planners; however, this group was squeezed in between the two other schools of thought who were pushing in the two opposite, extreme directions.

The other group was the radicals to whom the implementation of the investment ceiling was a first step in eventually eliminating private industrial investment except in the household and cottage industries. In their vision of the future economic system, the intermediate region between the state-owned large-scale industries, on the one hand, and the cottage and household industries, on the other, should progressively be owned and run either by the workers' co-operatives or the local government institutions or the agricultural co-operatives. This group strongly supported the maintenance of the investment ceiling because this would hopefully promote the ultimate objective of progressively reducing the scope of private investment through inflationary rise in the cost of investment.

Both these groups, i.e. the radicals and middle-of-the-road, were, however, opposed by those who wanted a greater role for private enterprise, beyond what was contemplated by the investment policy of 1972. According to this group, which included many members of the ruling party and of the bureaucracy as well as those who made fortunes in trading and speculative ventures, public enterprise should ultimately be restricted to only those areas where private enterprise was either unwilling or unable to invest. They were eager to ensure that private enterprise should receive liberal state patronage in terms of concessional credit, tax and foreign exchange facilities. In such a growth-oriented private enterprise system, inequalities and monopolistic or oligopolistic concentration of economic power were considered necessary costs of development. At the time of the revision of the industrial investment policy, no specific policies were contemplated which might act against the forces of concentration of economic fortunes and power. The argument advanced by this group that the investment ceiling was contributing to economic stagnation by frustrating private savings and investment as early as 1974 was largely spurious. Economic stagnation was due to a multiplicity of factors: it was difficult to quantify the separate influences of the various factors; moreover, empirical evidence in any case was not very conclusive or if conclusive, was not of the type which the political leadership could critically examine for the purposes of making up its mind. Because the political leadership was vacillating in its convictions and since economic distress in Bangladesh was partly due to inefficiency in management and organisation, for which it felt itself responsible, the persistent pressure by this powerful group against public enterprise and for a larger role of private investment had partly shaken the confidence of the leadership in its own policy.

It is difficult to predict the future pattern of economic system, including the role of private enterprise in Bangladesh. The socio-political developments in the last decade, as explained in the earlier chapters, which led to the enunciation of the investment policy in 1972 did not suddenly appear on the political scene in Bangladesh, nor would they necessarily cease to operate in future years. The radical forces undoubtedly suffered a temporary setback. Moreover, the moderate forces which opted for a mixed economy and evolutionary progress towards a socialist economy also had suffered a setback. It was unlikely that the trends on socio-political thinking which culminated in the economic policies immediately after independence would not reassert themselves. The complete and drastic reversal in policy initiated in the period subsequent to August 1975 indicated a state of unstable equilibrium between the contending forces; it would not be surprising if there was yet another reversal of policy in the opposite direction in the future.

Notes

1. Rehman Sobhan, "Nationalisation of Industries in Bangladesh: Background and Problems", *The Economic Development of Bangladesh within a Socialist Framework*, edited by E. A. G. Robinson and K. Griffin, Macmillan, 1974, pp. 181–5.

2. Ministry of Industry, *Industrial Investment Policy*, 1972/73, Preamble, July 1972. As the preamble to the statement on industrial investment policy in 1972 stated:

> The government in moving forward to deal with this problem [abandoned properties] recognised that this situation presented a historic opportunity to lay the foundations of socialism in the industrial sector of the economy. Socialism being one of the fundamental precepts of State policy, the government decided that when it had to assume responsibility for the bulk of the industrial sector, this should be affected within the total perspective which would identify the areas which should immediately be nationalised and taken into the public sector.

It further stated that the present phase was one "of consolidation of nationalisation, of stabilisation of new institutions and of development of a socialist consciousness among the management and workers".

3. There were 34 units in the jute industry and 24 units in the cotton textile industry which were under the ownership and management of Bangladesh entrepreneurs.

4. In Pakistani days, the private investors were allowed to have 80–90% debt and 10% equity ratio in their enterprises, debt being held by the industrial financing agencies. In addition, they had often indirect financing of their equity contributions through loans from commercial banks or from direct equity participation by state agencies so that private investors had contributed very little of their own savings or had undertaken very little of the risks; they mainly represented private management, excepting in the

case of joint enterprises with state participation, which frequently provided management as well.

5. Planning Commission, unpublished memorandum, July 1974.

6. At one stage, various alternatives were considered with respect to the disposal of the smaller enterprises such as industrial workers' co-operatives, agricultural co-operatives, commercial or industrial banks and local governments. It was finally decided to confine their disposal to private investors and workers' co-operatives, with a preference for the latter.

7. In 1970 a small-scale industry was defined as one which had assets not exceeding Tk.1 million; in view of the escalation of prices since 1970, the ceiling of Tk.2.5 million, for the period starting towards the end of 1972 onwards, would broadly refer to the same category of industrial units. Moreover, the limits of fixed assets below which the abandoned industrial units were to be sold out to the private entrepreneurs were placed at Tk.1.5 million in original book value, of which the replacement cost in prices of 1972/73 would have probably not exceeded 2.5 million.

8. In fact, public enterprises on the whole earned less than 15% return on investment in the last few years.

9. *The Five-Year Plan*, p. 39.

10. The Ministry of Finance wanted the maximum tax revenues or maximum surplus, the Ministry of Industry expected the maximum utilisation of capacity whereas the Finance and Foreign Trade Ministries made allocations of foreign exchange which were much less than what was required by the full capacity output; nonetheless the Ministry of Industry went ahead at the same time sanctioning additional capacity.

11. W. B. Reddaway, "Pricing Policy for State Enterprises in Bangladesh", *Bangladesh Development Studies*, January 1975. Also Planning Commission, *Annual Plan*, 1974/75, pp. 110–12, 140–2.

12. Ministry of Finance, *Budget Speech, Annual Budget*, 1971/72 and 1972/73, June 1972.

13. Government of Bangladesh, *Report of the National Pay Commission*, Vol. 1, Main Text, May 1973, pp. 21–2 and 159–62.

14. ibid,. p. 159.

15.

"In a socialist economic system not only the top executives should be committed to socialist goals but also they should be supplemented by the socialist cadres distributed within the middle management and amongst the workers, who would act as watchdogs over corruption and mismanagement and as motivating agents for higher productivity. In Bangladesh no such infrastructure was available but she had no alternative but to make the best use of the available professional and managerial skills even though they had no political commitment. They had to be motivated by material incentives, challenge of responsibility, a high social status and recognition". *Five-Year Plan*, p. 251.

16.

"The Liberation of Bangladesh has opened a new chapter in the economic and political life of the people. The Government is determined to root out all kinds of exploitation and remove social and economic inequality which prevailed in the country for so long. ... As a first step towards socialism the Government have already nationalised ... This is definitely a step forward in socialising our country and in improving the living and working conditions of our working class". *The Labour Policy*—Announcement on 27 September 1972 by the Minister of Labour.

17. Government of Bangladesh, *Report of the Industrial Workers' Wages Commission,* September 1973, p. 70.

18. N. Chowdhury, "Real Wages in Nationalised Sector in Bangladesh", Paper presented at the *Second Annual Conference of Bangladesh Economic Convention,* March 1976. A. R. Khan, *Poverty and Inequality in Rural Bangladesh,* ILO Working Paper, Geneva, March 1976, pp. 17–20.

19. Government of Bangladesh, *Labour Policy, Announcement by the Minister for Labour,* 19 September 1972.

20. When, therefore, in early January 1975 a one-party government was established, the multiple trade unions were also abolished.

21. *Labour policy,* August 1974. *Unpublished Note by the Ministry of Labour and Social Welfare,* Government of Bangladesh. *Report of the ILO/SIDA Mission on Workers' Participation in Management in Bangladesh,* ILO, Geneva, 1973.

22. That the middle classes were not likely to combine with the urban or rural proletariat to bring about a socialist revolution might or might not be true; it depended on the pattern and nature of future leadership. At the same time, it would not be said as the critics of the policy seemed to argue, that active encouragement to the development of large private capitalist interests in the industrial sector would have provided no obstacle to the future of socialist policies.

23. Government of Bangladesh, Ministry of Industries and Bangladesh Small Scale Industries Corporation. The average size of new industrial enterprises was about Tk.1 million in 1973/74 and Tk.1.84 million in 1974/75. The low average size of investment, in spite of the fact that costs of investment had increased since 1972 when the ceiling of Tk.2.5 million was fixed, contradicted the hypothesis of lack of viable investment opportunities within the ceiling.

24. Government of Bangladesh, *Priority List of Industries for the Years (1974–76) of Industrial Investment Schedule for the 1st Five-Year Plan* (1973–78), 1974.

25. IBRD, *Bangladesh: Development in Rural Economy,* Vol. 1, September 1974, p. 312.

26. Ministry of Industry, *New Investment Policy,* July 1974.

27. Ministry of Industry, *New Investment Policy,* June 1974.

28. Previously management contract was allowed in the case of collaboration between the public sector corporation and foreign private enterprise, the latter serving as the managing agent. Under the new policy, domestic private partners could also undertake a management contract in an enterprise jointly owned by public and private enterprise.

29. The public sector corporations purchased the abandoned units mainly for the sake of balancing the product composition or of assuring the supply of intermediate inputs produced by private units or of assuring an outlet for its products. In both cases, these units had a close interdependence with the public sector enterprises in terms of input–output flows.

30. Ministry of Industry, *Revised Industrial Policy,* December 1975. The industrial policy statement of 1975 contrasted with the policy of 1974, in not containing any reference to socialism. The 1974 statement even while

opening the door for an expanded private sector continued its formal rhetoric, one might say, of socialism, it stated,

> in pursuance of the government commitment to the establishment of a socialist economy, the key sectors including the bulk of the industries are already in the public sector. ... It is, however, considered conducive to the goal of rapid industrialisation and mobilisation of investment funds to afford scope to the private entrepreneurs to play an expanded role within the framework of a planned economy.

The 1975 statement abandoned the reference to socialism and justified the change in policy as follows:

> Government is of the view that while industries of basic and strategic importance or in the nature of public utility service, should remain in the public sector, for reaching the goal of mobilisation of investment funds and rapid industrialisation, the private entrepreneurs should be afforded adequate scope and facilities for setting up industrial enterprises within the broad framework of a planned economy, the provision of the constitution and the laws of the land and efforts should be made to consolidate and improve the efficiency of the public sector.

8. Conclusions

A number of lessons are to be learnt from the working of the planning machinery in Bangladesh. The most important is the need for participation by the political leadership in all stages of the planning process, starting from the broad outlines of a plan and its macro-economic implications to the details of preparation and implementation of programmes and projects. If the ministers are also engaged in planning exercises in their respective ministries, they would better understand the approach, constraints and possibilities faced in the overall planning framework. Since planning is a socio-political process, those who control and participate in the political process should not be kept out of the day-to-day deliberations in the planning process. If the technocrats are to provide the leadership of the Commission, they must be willing to play a political role; they must gain support amongst the interest groups and be ready to neutralise opponents, if they are to get their recommendations accepted and, more important, implemented. They not only have to understand political constraints on economic policies but also must analyse how the political feasibility constraints could be removed or contained within limits.

It is trite, but worth emphasising, that effective co-operation between the functional or administrative ministries and the Planning Commission is essential. Planning involves a constraint on their independence of action, especially of the functional ministries. If they actively participate in the deliberations of the Planning Commission so that planning decisions are a result of common exercise directed and supported by the political leadership, rather than dictates of a superior ministry, they may find compensation for their loss of freedom. In course of time in Bangladesh, the administrative ministries started to recognise the necessity of many of the planning functions. For example, it was gradually appreciated that there was a vital need for co-ordination in economic policy-making and implementation and that the planning machinery might as well be vested with some of the necessary tasks of co-ordination.

Another hard learned lesson is that if the planning machinery is run by "outsiders" and if the concept of an "elite" civil service in control of the governmental decision-making machinery dies hard, frictions can be avoided only by letting the elite civil service play a predominant role in the planning machinery; being all pervasive and entrenched in the government machinery, they can frustrate the planning machinery if they are bent upon it. On the other hand, tensions and strains as well as temporary dislocations which are inherent in any significant change in administrative structure may be an unavoidable price to pay for expanding the base of expertise and experience, including specialised skills, and for encouraging mobility between different services and professional groups. However, the price is worth paying only if the political leadership is determined to make the change; if the political leadership is indecisive and prefers to balance the competing groups in the administrative structure, it will set in motion an uncertain and defensive mode of action amongst them all; the result would be the worst of both worlds.

Any quantitative model is no substitute for judgement and supplementary analysis, especially in view of the inadequacy as well as the uncertainty of data and assumptions, as was the case of Bangladesh. Nevertheless, the interdependence between targets and instruments, between objectives and the corresponding policies and institutions required to achieve them need to be brought out clearly, if the politicians are to be able to take decisions wisely. Seldom, especially in Bangladesh, does the political leadership have a clearly defined preference function. Therefore, the Planning Commission does not have the task of simply working out the alternative policy instruments and strategies to achieve a given set of closely defined objectives or targets. When preferences are uncertain and changing, as is likely to be the case in most instances, the planners have the responsibility of working out the implications, in terms of policies and programmes, of alternative sets of objectives or preference functions. The need to do so reflects the fact that the political leadership is often unable to decide between objectives without knowing their policy implications. The alternative objectives do not merely include alternative rates of growth of income and consumption but also alternative socio-political and economic systems.

No Plan can remain unaffected by changes in underlying assumptions and exogenous variables such as technology, prices, investment alternatives, conditions of world trade, etc. The Annual Plan is a valuable instrument of planning not only in giving more concrete shape and content to a Five-Year Plan but also in adjusting plans to changes in assumptions and exogenous variables. No less important, as Bangladesh experience demonstrates, is the assumption regarding administration capacity and implementation ability. When drastic and violent changes take place in assumptions, they must be accommodated by changes in the long-term perspective of a Five-Year Plan.

Technological possibilities of increasing agricultural output, especially food production, are considerable in Bangladesh, but great uncertainty pertains to the effectiveness of the structure of incentives, management and institutions to realise these potentials. Stability in availability and prices of foodgrains is crucial to the success and stability of development efforts in a poor country like Bangladesh. Given considerable fluctuations in supplies and prices of foodgrains, one cannot over-emphasise the importance of building up stocks and reserves of foodgrains either from domestic production or from imports. In the face of endemic shortage it is crucial to build up stocks, if necessary by lowering the already meagre diet even in a good year. The dilemma, however, is that it is politically impossible to withdraw supplies either from the market or from the rationing system to build up stocks in a given year if supplies are not in excess of the minimum consumption requirements. Domestic procurement by the government often fails either because of administrative shortcomings or because unattractive prices are offered to the farmers. A remunerative price can only be ensured to the farmers if subsidies from the consumers are abolished in view of the lack of resources.

Experience with domestic resource mobilisation in Bangladesh re-emphasises the self-evident truth that the dominant groups, surplus farmers and traders, are not willing or ready either to bear an increase in the burden of taxation or a decline in the amount of subsidy. At a very low level of per capita income or consumption, indirect taxes which are good revenue earners are also those which fall on the widest number of consumers and, therefore, on the low-income groups. The fact that the rich were only few, does not reduce the need—indeed increases the rationale—of higher taxation on them, not because it would make significant additions to revenue but because it would make sacrifices on the part of the poor majority psychologically tolerable, and therefore, socially and politically acceptable. The fact that domestic savings impose sacrifices at the very low level of consumption in Bangladesh make it operative that investments are highly productive, i.e. efficiently managed and yield high returns. While the ability to impose increased taxes did face great obstacles, deficit financing in Bangladesh as a method of taxation caused less direct opposition because its effects were diffused and because the affluent farmers and traders generally gained from it. Inflation was a quiet way of raising resources for public investment without much publicity—unlike the proposals for increasing taxation or reducing subsidies in the budget. Its impact is felt only to the extent that groups adversely affected by inflation muster political support for agitation and public protests.

Admittedly it is both equitable and efficient to generate resources at the local level for investment in local projects, for example, for rural development. The process of decentralisation, however, faces a host of problems in Bangladesh and is easier said than done. The lack of viable local government institutions is an important hurdle in the way of development planning and its implementation. It was so in Bangladesh. A successful local government can seldom be a gift by the politicians of the national government. There must be popular pressure for the establishment of local government institutions. This is also linked with the spread of education and with an increase in general awareness among the masses about the desirability and possibility of their participation in the decision-making process of government. For a poor, resource-short country like Bangladesh, the cost of forgoing external assistance in terms of sacrifice of much-needed domestic consumption or development effort is too great. It is true that foreign aid does sometimes lead to a relaxation of domestic efforts either to improve efficiency or to raise domestic resources. Confronted with the need to increase output, often there is a strong temptation to add to capital equipment rather than to increase productivity, especially since the access to capital equipment is relatively easier given the donors' preference for project assistance. At the same time, it is also true that there are instances where pressures from the donors have led to domestic efforts to improve performance and rationalise policies.

The experience of Bangladesh also shows that the governments of many developing countries vigorously profess self-reliance but end up with large dependence on foreign aid, while at the same time finding such dependence politically embarrassing in terms of its image in the eyes of the public. This schizophrenic tendency is reflected in the government's forecasting less than the required foreign assistance in the Plans. This is partly due to an unrealistic assessment of the potentials of the economy or of the ability and willingness of the government to mobilise necessary domestic resources. To end up in the course of implementation of Plans with a desperate search for larger flow of assistance indicates weakness in a country's planning, as was the case in the Bangladesh Plan. Large dependence on foreign aid certainly introduces restrictions on the freedom of action in domestic and foreign policies, both economic and political, which can partly be reduced or offset by a wide diversification of the sources of foreign assistance as well as by refusing to accept foreign assistance, if, in the opinion of the political leaders, such constraints appear excessive. This requires that the leadership has sufficient self-confidence and mass support, and is able to motivate the public to suffer the necessary sacrifices in the interest of national integrity and independence.

The policies determining the relative incomes of different interest groups of the different sectors of the economy are no less important in a developing country than in an advanced country. To argue that the organised labour unions in the industrial sector should not press for higher wages because it increases inequality between them and the rural poor or urban unemployed, assumes that the surplus in the industrial sector is used to expand incomes and employment among the poor rather than being squandered on high consumption of the affluent or on unproductive projects; such an argument becomes especially indefensible in Bangladesh, when profit or property incomes remain unrestrained and inflation continues to eat up the real incomes of the industrial workers. A stable and disciplined labour force is essential for economic progress. Experience in Bangladesh reveals that industrial peace is difficult to achieve when the political process is volatile and rival political parties compete with each other to escalate workers' demands for better living conditions. To fix ceilings on salary incomes in the public sector as was done in Bangladesh, when incomes in the private sector are unrestrained, does not reduce inequity but may even accentuate it. At what point the low incomes of the public sector employees, especially in a period of inflation, increase the degree of corruption is a difficult question to answer. In a regime of discretionary controls a certain amount of corruption is unavoidable. It is doubtful whether corruption amongst public officials in Bangladesh was a significant source of misallocation of resources. Neither was it very effective in lubricating the wheels of government or the process of decision-making. A widespread public belief in corruption in high places in the government was, however, severely detrimental to the credibility of the government in the eyes of public, especially since the political leadership itself, in an unprecedented manner, proclaimed the prevalence of widespread corruption in the country. Corruption itself is bad enough; when its public admission was combined with inefficiency in administration and economic management it was used by opponents of the regime, as was the case in Bangladesh, both inside and outside the country, to seriously damage its credibility and integrity.

The debate on investment policy in the industrial sector seemed to indicate that mixed economies were subject to pressures from both the right and the left. The pressures were destabilising because the government and the party in power itself were unsure and did not seem to have firm convictions. This vacillation had the unfortunate result that whatever policies were decided upon were not implemented vigorously. Half-hearted policies inefficiently implemented produced the worst of both worlds. The historical circumstances leading to the creation of Bangladesh generated egalitarian expectations and reactions against large private fortunes and concentration of economic power. In this context, a firm commitment to a regime of untrammelled private enterprise is unlikely in the future. It would be far from easy to eliminate or stifle the socialist aspirations leading to pressures for public ownership and controls. It must be noted that it was not only policies but also socialistic pronouncements by the political parties and leadership which tended to dampen private investment. It would in all likelihood be difficult, with a view to promoting a thriving private sector, to introduce and effectively implement a pattern of financial and fiscal incentives and concessions, on a scale and for a period similar to that experienced in pre-independence Pakistan.

Private enterprise, whatever its magnitude or relative role, will continue to be hedged in with controls and regulations in a probably haphazard manner. There will continue to be shortages of all kinds for years to come and it is difficult to see, in the light of past experience, how shortages would not entail controls. Added to this would be a concern for equity and apprehension of a growth of monopolistic forces leading to a predilection for controls. It has never been cogently argued why a system of controls is necessarily superior to a system of public ownership. Weak and inefficient governments are a liability for any combination of public ownership and controls, on the one hand, and of private ownership and market economy, on the other.

There is indeed a dilemma. Those who have command over or access to economic resources or political decision-making power tend to gain cumulative advantage in the operation of free market forces or to strengthen their economic and political power by the manipulation of state patronage, controls and regulations in their favour. This tends to be true equally of the intermediate or middle classes like surplus farmers, traders, merchants, rentiers and middle class professionals, etc., as of the large-scale private industrialists or financial interests. The concept of a "neutral" state or a government run by a group of "declassed" visionaries holding the balance between the competing interest groups, but unrelated to and independent of them for its continuance in power, seemed not very realistic, even though it was to be devoutly wished. One can only hope that the enlightened self-interest of the dominant groups combined with pressures from organised but disadvantaged groups would lead to progress with equity in an evolutionary process.

Bangladesh is going to be a mixed economy in a state of flux for some years to come, subject to conflicting pressures as mentioned above. Under the circumstances, an important objective should be to concentrate on improvement in organisation and management, including efficiency of public sector enterprises, and on administration of controls in such a way as to reduce inefficiencies or inequities. Controls and interventions, if they tend to run significantly counter to the forces of demand and supply and to the need for incentives, would inevitably generate pressures for leakages or violations. A judicious combination of controls and prices, one reinforcing rather than frustrating the other, is a difficult but worthwhile objective to aim at.

It is in a way paradoxical that the debate on the future pattern of the economic system was conducted on and around the investment policy in the industrial sector, a very small fragment of the economy of Bangladesh. It is in the rural economy of Bangladesh that issues of private ownership, market economy, growth and equity and their interrelations would have to be debated and resolved. The mounting pressure of an increasing population on limited land area and the consequential growth of mini-farmers and landless labourers below the poverty line, however defined, are likely to bring these issues to the fore. That many urban middle classes are linked up with landed interests and have a share in income from land, complicates the nature of interaction between pressure groups, linking changes in the rural economy with those in the urban sector, and increasing the uncertainty of outcome.

At least in the immediate future, the broad outlines of a policy package for rural development seem to lie in the following direction: a redistribution of land, a large increase in the supply of agricultural inputs and raw materials for rural industries, an elastic supply of credit, extensive rural education and training, combined with appropriate price signals and technology for efficient resource allocation and incentives. They all involve difficult economic, political and organisational problems. While they seem to be, at least in light of the existing state of knowledge, policies for non-devolutionary progress in rural Bangladesh, they also raise a number of questions. What are the political preconditions for such a change and what combination of interest groups would generate it? Would the enlightened self-interest, conceived with a sufficiently long time horizon, of the urban middle classes and intelligentsia succeed in galvanising sufficient political strength to organise the rural poor, to take measures for their protection and increased economic strength, to curb or overcome the opposition of the surplus farmers and mercantile interests in the rural sectors and to implement such a policy package? What would be the time horizon for the implementation of such a policy? These questions are obviously the subject of a different and more profound study.

Index